Olive Cultivation

Edited by Taner Yonar

Published in London, United Kingdom

Olive Cultivation
http://dx.doi.org/10.5772/intechopen.98132
Edited by Taner Yonar

Contributors
Theofilos Frangopoulos, Stefano Grego, Giora Ben-Ari, Iris Biton, Dvora Namdar, Yair Mani, Fatemeh Rahimianfar, El Hassan Sakar, Said Gharby, Annette Bongartz, Martin Popp, Richard Retsch, Dietrich Oberg, Karima Bouhafa, Valentina Giovenzana, Alessia Pampuri, Roberto Beghi, Andrea Casson, Riccardo Guidetti, Alessio Tugnolo

First published in London, United Kingdom, 2022 by IntechOpen
IntechOpen is the global imprint of INTECHOPEN LIMITED, registered in England and Wales, registration number: 11086078, 5 Princes Gate Court, London, SW7 2QJ, United Kingdom

British Library Cataloguing-in-Publication Data
A catalogue record for this book is available from the British Library

Additional hard and PDF copies can be obtained from orders@intechopen.com

Olive Cultivation
Edited by Taner Yonar
p. cm.
Print ISBN 978-1-80355-441-9
Online ISBN 978-1-80355-442-6
eBook (PDF) ISBN 978-1-80355-443-3

Meet the editor

Prof. Dr. Taner Yonar is a professor in the Engineering Faculty, Environmental Engineering Department, Bursa Uludağ University, Turkey, where he received his BSc in Environmental Engineering in 1996 and MSc and Ph.D. in Environmental Technology in 1999 and 2005, respectively. He completed his post-doctoral research in the Chemical Engineering and Advanced Materials Department, Newcastle University, United Kingdom in 2011. He teaches graduate- and undergraduate-level courses on water and wastewater treatment and advanced treatment technologies. He has been working to transfer his academic experience to the industry with Envora R&D Engineering Co., which began as a start-up initiative in 2020. He works on advanced oxidation technologies, membrane processes, and electrochemical processes. Dr. Yonar is the editor of five books and the author of more than eighty research papers.

Contents

Preface

Olive is a plant that has served as a source of food and even healing for thousands of years, especially in the Mediterranean Basin. It is also the source of livelihood for many people in the countries surrounding the Mediterranean Sea. This book presents a comprehensive overview of the olive plant in nine chapters written by experts in the field.

The first section of the book includes three chapters. In Chapter 1, Dr. Bouhafa discusses olive tree fertilization management in Morocco as a case study. In Chapter 2, Prof. Grego gives a detailed history and discusses the importance of olive. In Chapter 3, Dr. Giovenzana et al. discuss optimization techniques of the olive production chain using optical techniques and the development of new cost-effective optical systems inspired by agriculture 4.0.

The second section of the book includes six chapters. In Chapter 4, Dr. Frangopoulos evaluates the effects on color stability of incorporation of Trigonella Foenum-Graecum seed powder in nitrite-free meat emulsion systems with olive oil. In chapter 5, Mrs. Bongartz et al. discuss whether different test locations have a relevant impact on data quality during the sensory evaluation of extra virgin olive oil. In Chapter 6, Dr. Ben-Ari et al. review published studies regarding the mechanical harvest of table olives and attempt to identify the main issues in this process. In Chapter 7, Prof. Sakar and Prof. Gharby examine extraction technologies, chemical composition, and enrichment techniques by natural additives on olive oil. In Chapter 8, Dr. Oberg proposes additional sensory criteria for the quality assessment of the grade of extra virgin olive oil. Finally, in Chapter 9, Dr. Rahimianfar discusses the effect of olive leaf extract on systolic and diastolic blood pressure in adults.

Taner Yonar
Environmental Engineering Department
Bursa Uludağ University,
Bursa, Turkey

Section 1

Olive as a Plant

Chapter 1

Management of Olive Tree Fertilization in Morocco

Karima Bouhafa

Abstract

This chapter focuses on olive tree fertilization in Morocco: Describe the practices used by olive growers, diagnose the nutritional status of olive orchards and synthesize the different results and the recommendations of research carried out in Morocco around this theme. Before that, a general overview of the olive tree nutrition and its needs in mineral elements as well as the role of each of these nutrients in the olive tree growth and development will be presented. An introduction to the importance given to the olive tree in national agricultural strategies is necessary. The surveys carried out in the Sais region have shown that farmers do not control olive tree fertilization. This affected negatively the soil fertility level and the olive tree's nutritional status, which were determined through soil and leaf analyzes. From the results of three field trials, carried out in the Fez-Meknes region, it can be concluded that nitrogen and potassium are the two most important elements for the olive tree nutrition and which can affect both its productivity and its quality. The impact of phosphorus on the crop has not been significant, whereas our farmers provide it in high doses compared to the crop's need.

Keywords: Olea Europea, Morocco, fertilization, macronutrients, surveys, experimentations

1. Introduction

With an area of around 1.17 million hectares, 60% of which is cultivated in rainfed conditions, the olive tree occupies a preponderant place in the national arboreal sector. It plays an important role in promoting the economy and employment as it contributes 5% to the agricultural gross domestic product (GDPA) and generates around 100,000 permanent jobs. Despite the continuous increase in its area, in response to the State's strategy (Green Morocco Plan) to reach 1.22 M ha by 2020, its production remains low and below the potential of the sector, especially in the rainfed system where yields in olives rarely exceed 3 $T.ha^{-1}$. This low yield is the result of two main factors, which are as follows:

- The climate and in this case drought.
- Faulty technical management of the crop, including fertilization, pruning and diseases, and pests incidence.

Fertilization, the subject of our chapter, is a very important cultivation technique for all agricultural production; it provides the crop with the nutrient requirements necessary for its growth and development. However, the majority of Moroccan olive growers, especially small farmers, consider the olive tree to be a hardy species that does not require maintenance. Also, the absence of fertilization standards for the olive tree, adapted to each agro-climatic region, leaves Moroccan olive growers with the obligation to follow traditional fertilization practices or in the best cases to fall back on recommendations obtained in other countries. Hence, the fertilizers brought by the Moroccan olive growers are, in the majority of cases, random both for quantity and quality, because it is not based on soil or vegetal analyzes.

All this prompted us to ask the following questions:

- Are our olive orchards well-nourished with essential elements to be able to ensure good productivity?
- Does fertilization significantly affect olive production?
- How should we reason the fertilization of the olive tree?

2. Assessment of soil fertility and nutritional status in olive orchards

We proceeded with a diagnostic study of soil fertility and olive tree nutritional status in 58 orchards chosen at random in Central Morocco. This study was carried out through soil analysis, surveys, and leaf analysis. Composite soil samples were, therefore, taken from the two layers of 0–30 and 30–60 cm during the period of vegetative rest of the crop, which is December–January. Leaf samples were also taken during the same period and from the same olive orchards.

2.1 Description of studied olive orchards

Through the surveys carried out, we have observed that the Moroccan Picholine variety dominates (98.3%) in olive orchards and that the olive tree is associated with intercropping in almost half (48.3%) of the orchards studied. The chosen sample is characterized by different age categories ranging from 4 years to over 70 years. Orchards belonging to the age group between 11 and 40 years old represented 60% of all the olive groves surveyed, while young orchards whose age does not exceed 10 years represented only 5%. More than half (57%) of the orchards studied are managed in rainy conditions. In irrigated orchards, the gravity irrigation system dominates with a proportion of 24% against 19% of all orchards surveyed for the drip irrigation system. The majority of plantations (85%) have a planting density between 100 and 350 trees.ha^{-1}. We noted that about 30% of the orchards studied have a planting density between 200 and 350 trees.ha^{-1}. This is the optimal density recommended for the Moroccan Picholine variety in the study area. However, the study showed lower densities ranging from 100 to 200 trees.ha^{-1} at 55% of the olive groves studied. The yields declared by the olive growers surveyed varied greatly from one farm to another. They varied between 0 and 14.3 T.ha^{-1}. The average olive yield was higher (5.1 T.ha^{-1}) in olive trees under drip irrigation system, compared to those under gravity irrigation system (1.9 T.ha^{-1}) and in rainy conditions (1.8 T.ha^{-1}). Olive orchards whose planting density belongs to the density class [200–350] trees.ha^{-1}, which represents the optimum planting density for the Moroccan variety Picholine at the regional level, achieved the best average olive yield (3.7 T.ha^{-1}), in comparison with the other density classes identified. On the other hand, the extensive densities (<100 trees.ha^{-1}) allowed the minimum average yield (1.5 T.ha^{-1}).

2.2 Fertilization practices adopted by olive growers

About 48% of the olive growers surveyed do not use any mineral fertilizers for their olive trees. The absence of fertilization had repercussions of course on the olive yield that was on average 3 T.ha^{-1} in the fertilized orchards against 1.9 T.ha^{-1} achieved in the unfertilized olive orchards (**Table 1**).

The calculation of the average doses of nitrogen, phosphorus, and potassium in the orchards studied showed that nitrogen (N) is the most provided element by farmers with an average dose of 37 Kg N.ha^{-1}, followed by phosphorus (P) with 20 Kg P_2O_5.ha^{-1} and finally by potassium (K) with 9 Kg K_2O.ha^{-1} (**Table 2**).

If we consider the most abundant density in our sample, which is 100 trees.ha^{-1} (42.4% of orchards), these average doses applied become as follows—0.37 Kg N.$tree^{-1}$, 0.2 Kg P_2O_5.$tree^{-1}$, and 0.09 Kg K_2O.$tree^{-1}$. And these are low doses for the olive tree. In addition, phosphorus is supplied by some farmers at very high doses that even exceed nitrogen and that have reached 138 Kg P_2O_5.ha^{-1}. The times of fertilizer input were generally concentrated over the period from January to April indicating an absence of inputs during other periods where the need for mineral elements is important for the olive tree, such as the fruit growth phase.

Finally, all these data indicate the existence of a failure in the fertilization practices adopted by the olive growers surveyed concerning both the fertilizer doses applied and their application moments.

	Fertilized orchards	Unfertilized orchards
Average yield (T.ha-1)	3	1.9

Table 1.
Average olive yields.

	Nitrogen (Kg N.ha-1)	Phosphorus (Kg P2O5.ha-1)	Potassium (Kg K2O.ha-1)
Minimum	0	0	0
Maximum	77	138	50
Mean	37	20	9

Table 2.
The nitrogen, phosphorus, and potassium quantities provided by olive growers surveyed.

2.3 Assessment of soil fertility

Analyses of soil samples taken from the orchards have shown that the majority of these soils are basic, limestone, and largely poor to moderately provided with organic matter (**Table 3**).

	pH		% active limestone		% organic matter	
	0–30 cm	30–60 cm	0–30 cm	30–60 cm	0–30 cm	30–60 cm
Minimum	6.3	6.4	0	0	0.1	0.04
Maximum	8.8	8.9	18	1[illegible].8	4.5	2.7
Mean	8.0	8.2	12.5	1[illegible].9	2.1	1.3

Table 3.
Soil sites characterization.

Previous studies have shown that the olive tree tolerates a wide pH margin, but values between 7 and 8.5 allow its best development [1]. Other studies have also shown that excellent yield and vegetative growth can exist on olive grove soils with low limestone content and 50% limestone [2]. Therefore, the pH and % limestone of the studied soils are favorable for good growth and good development of the olive tree. We found that the average soil organic matter content was higher in olive orchards associated with intercropping compared to those conducted in monoculture (**Table 4**).

These results could be explained by the residues of these crops associated with the olive tree that certainly contributed to a greater accumulation of organic matter in the soil.

Soil analysis results showed that soil nitrate contents varied between 1.8 and 71.4 $mg.Kg^{-1}$ and between 1.5 and 40 $mg.Kg^{-1}$, respectively, for the 0–30 cm and 30–60 cm layers. For available phosphorus, the soils presented contents ranging from 1.3 to 59.3 $mg.Kg^{-1}$ for the 0–30 cm layer and from 1.4 to 41.7 $mg.Kg^{-1}$ for the 30–60 cm layer. Exchangeable soil potassium fluctuated between 43.8 and 1456.5 $mg.kg^{-1}$ for the 0–30 cm layer and between 34.4 and 997.7 $mg.kg^{-1}$ for the 30–60 cm layer. According to the interpretation standards for soil analyses defined by the California Fertilizer Association [3], 50% and 84.5% of the studied soils are poor in phosphorus, respectively, for the 0–30 cm and 30–60 cm layers. In contrast, soil potassium levels were low to medium in 15.5% and 55% of olive orchards, respectively, for soil layers 0–30 cm and 30–60 cm.

These results confirmed the existence of a deficiency in the fertilization practices adopted by the farmers.

2.4 Evaluation of the nutrient state of olive orchards

Olive leaf analysis revealed low levels of N, P, and K that varied, respectively, from 0.22 to 0.60%, from 0.04 to 0.26%, and from 0.34 to 1.08%. The results showed that leaf macro elements levels were, in the majority of cases, below the deficiency thresholds cited in the literature [4]. In fact, all of the orchards studied require nitrogen inputs and almost 91% of the orchards need potassium fertilization. As for phosphorus, it caused less problems compared to nitrogen and potassium since only a third of the orchards sampled required phosphorus input.

Soil layer	0–30 cm	30–60 cm
Olive tree without intercropping	1.9%	1.2%
Olive tree with intercropping	2.4%	1.4%

Table 4.
Soil organic matter in intercropping system.

Yield	Rainy conditions		Gravity irrigation		Drip irrigation	
	Model	R^2	**Model**	R^2	**Model**	R^2
Nitrates	y = −0.0005x2 + 0.0257x + 1.5725	R^2 = 0.01	y = 2.96e-0.036x	R^2 = 0.30	y = −0.0007x2-0.0638x + 6.6668	R^2 = 0.08
Available P	y = 0.0002x2-0.0365x + 2.3283	R^2 = 0.03	y = −0.0032x2 + 0.1696x + 0.3605	R^2 = 0.18	y = −0.017x2 + 0.9779x-3.967	R^2 = 0.43
Exchangeable K	y = 1E-06x2-0.0013x + 2.0524	R^2 = 0.01	y = −1E-05x2 + 0.0087x + 0.4994	R^2 = 0.12	y = 0.0002x2-0.094x + 17.129	R^2 = 0.17

Table 5.
Regressions between olive yield and soil nitrate, available phosphorus, and exchangeable potassium content, according to the three water regimes of olive groves studied.

Yield	Rainy conditions		Gravity irrigation		Drip irrigation	
	Model	R^2	Model	R^2	Model	R^2
% N	y = −8.8455x2 + 8.6368x-0.1061	$R^2 = 0.01$	y = 184.75x2-118.7x + 20.193	$R^2 = 0.35$	y = −166.6x2 + 125.2x - 17.805	$R^2 = 0.03$
% P	y = 193.49x2-38.882x + 3.4603	$R^2 = 0.03$	y = 300.44x2-68.193x + 4.9764	$R^2 = 0.63$	y = −720.24x2 + 253.39x - 13.875	$R^2 = 0.16$
% K	y = −12.073x2 + 16.425x - 3.3019	$R^2 = 0.03$	y = 4.2404e-1.676x	$R^2 = 0.32$	y = −185.75x2 + 201.21x - 46.614	$R^2 = 0.24$

Table 6.
Regressions between olive yield and leaf macroelements contents, according to the three water regimes of olive groves studied.

2.5 Correlation between parameters studied

We looked for correlations between the olive yield and the soil contents of nitrates, available P, and exchangeable K (**Table 5**) on the one hand, and between the olive yield and the leaf contents of N, P, and K (**Table 6**) on the other hand. But none of these correlations have been confirmed for the three existing water regimes (rainy conditions, gravity, and drip irrigation system).

A study carried out in Syria revealed that the olive yield variability was explained at 68% by the amount of potassium available in the root zone, followed by total N with 58% and mineral N with 44% [5]. The same study showed the absence of correlation between yield and leaf N and P contents and a significant correlation (26%) between yield and leaf K content.

We also studied the relation between olive leaf nitrogen, phosphorus, and potassium contents and soil nitrates, available phosphorus, and exchangeable potassium contents at the 0–30 cm layer, always for each water regime adopted by farmers (rainy conditions, gravity, and drip irrigation system). The results obtained showed that olive nutrition parameters are not linked to soil fertility parameters in these orchards. The same result was reported by a study carried out in Tunisia for P and K [6].

3. Olive tree fertilization

3.1 Synthesis of research work carried out in the Mediterranean basin

Mineral nutrition is one of the major factors in optimizing fruit yield and quality [7]. For the olive tree, nitrogen (N), phosphorus (P), and potassium (K) are essential nutrients. Marin and Fernández-Escobar [4] reported that annual intake is not necessary for good olive productivity. Hence, technical management can be inefficient following an underestimation or an overestimation of inputs at the orchard level. In fact, under-fertilized areas do not reach optimum yield levels, whereas, in over-fertilized areas, there could be a high risk of environmental pollution and an increase in costs [8]. Centeno and Gómez del Campo [9] reported an increase in olive yield after N application to the soil and P and K application by foliar spraying, although initial leaf analyses indicated adequate nutrition levels. After five trial years, Fernandez-Escobar et al. [10] reported that when olive tree fertilization is based on foliar diagnosis, it satisfies crop nutrient needs, minimizes environmental impact, improves crop quality, and avoids excessive and systematic use of fertilizer. In Spain, Garcia [11] proposed, for the olive tree, a balanced formula between the macro elements of 20-8-14 (N-P_2O_5-K_2O) based mainly on the olive's nutrients exports.

3.1.1 Nitrogen

A survey carried out across the Mediterranean basin where about 98% of the 10 million hectares of existing olive groves in the world are located [12], showed that nitrogen is present in most fertilizer applications, even when potassium is the element that causes most severe nutritional disorders [13]. In a long-term experiment conducted with rain-fed olive orchards in several localities in Spain, Ferreira et al. [14] found that only trees with productivity below 35 $kg.tree^{-1}$ showed a positive response to N intake. For a period of experimentation of 13 years, Fernández-Escobar et al. [15] found that nitrogen fertilization did not have significant effects on yield, fruit characteristics, and tree growth in two typical orchards of the Mediterranean region so leaf nitrogen concentration increased with nitrogen dose. They also noted the absence of a yield decrease or olive tree growth decrease even when leaf N content was below the established deficiency threshold (1.4%), thus suggesting that this deficiency threshold should be inferior. A combination of soil N inputs and foliar N applications (50% to soil and 50% foliar) was more effective in increasing the olive leaf nitrogen, compared to the supply of the totality of N to the soil; this can reduce the amount of nitrogen fertilizer needed to correct a possible N deficiency [15]. Rodrigues et al. [16] reported a gradual and significant decrease in olive yield when nitrogen was removed from the fertilization plan for 4 years, compared to treatments where nitrogen was added annually. Jasrotia et al. [17] also found a significant increase in olive tree productivity with increasing nitrogen doses. After 5 years of study in olive orchards in southern Spain, Fernández-Escobar et al. [18] found no significant differences in terms of olive yield between trees subjected to a fertilization program based on foliar diagnosis, and those receiving, annually, the current fertilization in this region (500 $kg.ha^{-1}$ of an NPK fertilizer (15-15-15) plus three foliar sprays of trace elements and amino acids). They also found that traditional fertilization practiced by farmers increased fertilization cost by more than ten times without increasing yield, vegetative growth, or oil content. In addition, the excess nitrogen affected negatively olive quality by inducing a decrease in polyphenol content with important antioxidant effects for olive oil. Nitrogen promotes an increase in the oleic and stearic acid contents of drupe and its deficiency is accompanied by an increase in palmitic and linoleic acid levels [19]. Too much nitrogen can cause environmental degradation [20] and affect negatively the groundwater quality [21]. This excess of N can also affect the olive oil quality [22] and the flower quality by reducing the egg's longevity [13]. These latter authors have also shown that a nitrogen deficiency caused an increase in the pistil abortion for the olive tree but only during the year when rainfall was low during the period preceding flowering. They suggested that a pre-flowering water deficit coupled with nitrogen deficiency induces an increase in pistil abortion for olive trees.

3.1.2 Phosphorus

Generally, phosphate fertilization is not recommended or practiced in rain-fed olive orchards [23]. Several authors have tried to determine the limiting and optimal values of soil available phosphorus concentration. Gargouri and Mhiri [6] found a critical value of 8 mg $P.kg^{-1}$ obtained by the Olsen method. Previous work has shown that responses to phosphorus are rare in fruit trees [24] and it has not been clearly demonstrated for olive trees in the field [25, 26]. Rodrigues et al. [27] suggested that regular intakes of P might not be necessary, in agreement with other opinions [25, 26]. They also reported that the low level of olive phosphorus exports may explain the crop's lack of response to P fertilizer inputs observed in field trials. In contrast, Fontanazza [28] reported that phosphorus deficiency limits the absorption of nitrogen, magnesium, calcium, and boron and consequently reduces plant growth.

3.1.3 Potassium

Potassium fertilization is considered essential for the olive tree, especially because the fruit is highly concentrated in K [29]. Ben Mimoun et al. [30] reported a positive effect of potassium fertilization on olive yield and oil content under rainy conditions. Potassium is known not only for its significant effect on yield and fruit quality but also for its effect on water use efficiency [31]. Adequate potassium fertilization allows better tolerance to a drought season [32], which is very common under our Mediterranean conditions [21]. In their study, Ben Mimoun et al. [30] found that fractional foliar potassium inputs had a greater effect than soil potassium inputs on the olive tree. This implies that this technique is preferable especially under rainy conditions because the lack of moisture in the soil during the plant's growth period could limit potassium absorption by the roots. Nutrient uptake depends on the supply of nutrients to the root system, namely their availability, the nutrient requirement level, and the absorption period [33]. Fine-textured soils are characterized by potassium uptake, so the addition of potassium to the soil surface is almost ineffective [34]. Foliar nutrient inputs are, in general, useful for meeting plant requirements and have high efficiency [35]. Potassium is particularly well suited to this fertilization form because just after foliar spraying, its translocation takes place quickly through the leaves [34]. The minimum threshold of the soil's available potassium content correlated with its clay content. This threshold is 80 mg $K.Kg^{-1}$ when the clay percentage is less than 15% and 150 mg $K.Kg^{-1}$ beyond this limit [6]. These potassium thresholds were obtained by the K extraction method with ammonium acetate. Sarrwy et al. [36] reported a remarkable improvement in leaf nutritional status, yield, and fruit quality after the application of potassium nitrate or mono-potassium phosphate, compared to control trees. The best result has been obtained with potassium nitrate, which is probably due to the high nitrogen requirement in the olive tree nutrition compared to phosphorus. Garcia [11] recommended 1 to 2 Kg $K_2O.tree^{-1}$, based mainly on the exports of olives in potassium. Although potassium is often a nutritional problem in olive orchards [25, 26], high doses of fertilizer may not be necessary [27]. The need for a regular supply of potassium, and the dose that must be provided for each application, depends on the availability of K in the soil and on the latter's capacity to retain it adsorbed by colloids or fixed by clay minerals. In sandy soils, for example, the strategy for supplying K should be similar to that for N, based on a regular supply of a limited amount of fertilizer. In clay soils, it is possible to provide higher doses with less frequent applications [27].

From this literature review, we can say that nitrogen and potassium are the most elements required by the olive tree, compared to phosphorus, which poses fewer problems.

3.2 Some results on olive tree fertilization in Morocco

In Morocco, studies on olive tree fertilization are almost non-existent. Few studies have looked at this aspect. Generally, the olive tree is considered, especially by small farmers, as a hardy species that does not require maintenance. As a result, determining the fertilization standards for the olive tree is essential for the rationalization of fertilizer inputs, in particular nitrogen, phosphates, and potassium. These macro elements are generally the most required by olive trees and will help improve crop yield levels. In this part, we will report some results related to three olive fertilization trials.

Three trials were carried out in rainy conditions in 3 different sites (S1 = Taza, S2 = Taounate, and S3 = Fez) belonging to the Fez-Meknes region which encompasses 33% of the total national olive tree area. Two orchards among the three chosen are planted by the Moroccan variety Picholine that dominates in Morocco. For the same variety, we considered two different age categories: a young orchard (S1: 9 years old) and an old orchard (S2: 35 years old) but with, nearly, similar planting densities (10 * 10 for S1 and 9 * 9 for S2). The third site was represented by a young orchard (S3) of the Spanish variety Arbequina with a higher planting density (3 * 5).

Before the installation of these experiments, a soil physicochemical characterization in the three sites was carried out through laboratory analyzes of the soil samples taken from two soil layers 0–30 and 30–60 cm. The analysis results showed that the studied soils are basic, poor in organic matter, non-saline for olive trees, and moderately to strongly calcareous for S1 and S3 and non-calcareous for S2 (**Table 7**). We noted low soil available phosphorus content at S1 and S2 and lower soil nitrate content at S2. For exchangeable potassium, these soils were well provided with this element [3].

The study design adopted for these trials is factorial in incomplete random blocks with two blocks. Each elementary plot consists of four trees. Four doses of each of the elements N, P, and K were tested. The nitrogen was fractionated into two inputs—1/2 in March and 1/2 in May. Phosphorus and potassium were brought in March.

At S1 and S3, the olive yields were equal, while the planting densities, as well as the olive tree varieties, were different. Generally, an olive tree in a dense orchard (S3) would produce less compared to another in an orchard where planting density is low (S1). In the latter case, competition between trees for nutrients, water, and light is weak. The yield recorded at the S3 level can be explained on the one hand, by the significant amount of rain (797.4 mm) that it received during this year compared to S1 (580.1 mm) and S2 (499.6 mm), and on the other hand, by the significant production potential of the Arbequina variety planted in this orchard. This potential was proved by a study in Tunisia where the behavior of different introduced varieties and Tunisian varieties was studied, the evaluation of the production potential of these varieties showed that Arbequine comes in the first position next to the variety Chemlali about cumulative production [37].

3.2.1 Nitrogen

The result showed that at S1 (9 years) and S3 (7 years), nitrogen input was not necessary since it did not improve the productivity parameters of these olive orchards (**Table 8**) and negatively affected the olive oil quality, especially peroxide index (**Table 9**). This could be due to the availability of soil mineral nitrogen, needed by olive trees in these orchards (**Table 7**). On the other hand, at S2 (35 years old), the addition of nitrogen fertilizer was beneficial since it improved both yields, yield efficiency as well as olive oil content. In the latter site, the nitrogen requirement of the olive tree was relatively high given

Site	**Taza (S1)**		**Taounate (S2)**		**Fez (S3)**	
Depth (cm)	**0–30**	**30–60**	**0–30**	**30–60**	**0–30**	**30–60**
Texture	Loam	Loam	Silty clay	Silty	Silty	Siltyclay
pH	7.9	7.6	7.6	7.8	7.5	7.9
Electrical conductivity (dS.m^{-1})	0.687	2.95	0.248	0.24	1.745	1.365
Nitrates (mg.Kg^{-1})	17.1	40.7	9.5	8.4	47.2	16.4
Available P (mg P.Kg^{-1})	5.7	5.2	6.7	1.5	33.6	29.1
Exchangeable K (mg K.Kg^{-1})	541.3	319.2	408.6	233.6	318.5	142.4
Organic matter (%)	1.5	2.2	2.3	1.8	2.3	
Total limestone (%)	20.9	24.7	2.2	1.6	19.6	
Active limestone (%)	10.3	9.9	—	—	9.81	

Table 7.
Physicochemical characteristics of the experimentation soils sites.

Kg N.tree^{-1}.year^{-1}	Yield (Kg.tree^{-1})			Yield efficiency (Kg.cm^{-2})		
	S1	S2	S3	S1	S2	S3
0	64.4a	10.7c	53.4a	0.09a	0.01b	0.85a
0.25	45.3a	25.1b	57.2a	0.10a	0.01b	0.74a
0.5	53.6a	58.1a	58.1a	0.11a	0.03a	0.79a
1	48.5a	21.9b	52.7a	0.09a	0.01b	0.76a

Table 8.
Nitrogen effect on olive yield.

	Site 1			Site 2	Site 3		
	Oil	Acidity	Peroxide index	Oil	Oil	Acidity	Peroxide index
Kg N.tree^{-1}.year^{-1}	%	%	Meq O_2.Kg^{-1}	%	%	%	Meq O_2.Kg^{-1}
0	25.3a	0.81a	3.5b	14.0c	38.5a	4.16a	7.8b
0.25	29.7a	0.74a	13.4ab	15.6c	37.0bc	3.11ab	7.1b
0.5	29.4a	0.76a	17.3a	21.1b	37.3b	2.51b	14.8ab
1	28.8a	1.67a	6.4ab	31.0a	36.3c	3.04ab	22.6a
Mean	28.3	1.00	10.2	20.4	37.3	3.21	13.1

Table 9.
Nitrogen effect on olive oil content and quality.

its age in comparison with the other two young orchards. This high nitrogen requirement, combined with initial low nitrogen content in the soil (**Table 7**) could explain this response of the olive tree to nitrogen at S2. The low yield at S2 could also be the result of a lack of water during the period before flowering and which negatively affected the latter. The 0.5 Kg N.tree^{-1}.year^{-1} dose allowed the best yield (57.7 Kg.tree^{-1}). This result found at S2 is in agreement with that reported by Garcia [11] who recommended 0.5 to 1 Kg N.tree^{-1} based mainly on the nitrogen exports of the olive tree.

3.2.2 Phosphorus

Phosphorus did not have a significant impact on the Olive tree productivity and quality at the three test sites (**Tables 10** and **11**). These results confirm previous research that suggested that regular phosphorus intakes may not be necessary [13, 26, 27] and that phosphorus is not generally recommended in rain-fed olive orchards [23].

Kg P_2O_5.tree^{-1}.year^{-1}	Yield (Kg.tree^{-1})			Yield efficiency (Kg.cm-^{2})		
	S1	S2	S3	S1	S2	S3
0	47.1a	20.8a	54.4a	0.09a	0.01a	0.75a
0.12	53.4a	28.9a	57.0a	0.11a	0.02a	0.79a
0.25	55.6a	28.4a	55.6a	0.08a	0.02a	0.81a
5	54.4a	26.1a	54.4a	0.10a	0.01a	0.79a

Table 10.
Phosphorus effect on olive yield.

	Site 1			Site 2	Site 3		
	Oil	Acidity	Peroxide index	Oil	Oil	Acidity	Peroxide index
Kg P_2O_5.tree^{-1}.year^{-1}	%	%	Meq O_2.Kg^{-1}	%	%	%	Meq O_2.Kg^{-1}
0	27.8a	0.71a	9.2a	19.7a	37.2bc	3.39a	13.5a
0.12	27.2a	0.76a	6.1a	20.8a	37.3b	3.02a	9.3a
0.25	30.3a	0.70a	11.7a	19.9a	36.4c	2.53a	14.1a
0.5	27.8a	1.81a	12.4a	21.3a	38.4a	3.88a	15.4a
Mean	28.3	1.00	10.2	20.4	37.3	3.21	13.1

Table 11.
Phosphorus effect on olive oil content and quality.

3.2.3 Potassium

The effect of potassium on yield and yield efficiency appears only at S3 (**Table 12**), but it did not affect either olive oil content or quality (**Table 13**). The non-response of olive trees to potassium input at S1 and S2 could be due to soil potassium richness in these orchards. This response of the olive tree to potassium supply at S3 may be due to the relatively low soil potassium content in comparison with S1 and S2 if we refer to the soil clay content which made potassium unavailable for the crop. An input of 0.5 Kg K_2O.tree^{-1}.year^{-1} was, therefore, necessary and sufficient to improve olive yield at this site. While in Spain, Garcia [11] recommended 1 to 2 Kg K_2O.tree^{-1} based on olive tree's exports of potassium.

Kg K_2O.tree^{-1}.year^{-1}	Yield (Kg.tree^{-1})			Yield efficiency (Kg.cm^{-2})		
	S1	S2	S3	S1	S2	S3
0	44.4a	29.2a	50.6b	0.11a	0.05a	0.76b
0.5	53.9a	25.1a	56.7a	0.11a	0.01a	0.79a
1	52.8a	29.0a	56.6a	0.08a	0.02a	0.78a
2	60.6a	21a	57.5a	0.09a	0.01a	0.81a

Table 12.
Potassium effect on olive yield.

	Site 1			Site 2	Site 3		
	Oil	Acidity	Peroxide index	Oil	Oil	Acidity	Peroxide index
Kg K_2O.tree^{-1}.year^{-1}	%	%	Meq O_2/kg	%	%	%	Meq O_2/kg
0	27.7a	0.76a	9.4a	19.4a	37.7a	3.13a	14.9a
0.5	28.2a	0.70a	5.3a	21.4a	37.3a	2.78a	9.7a
1	29.2a	0.78a	11.8a	21.1a	36.5a	3.60a	11.4a
2	28.1a	1.75a	12.7a	20.0a	37.7a	3.32a	16.3a
Mean	28.3	1.00	10.2	20.4	37.3	3.21	13.1

Table 13.
Potassium effect on olive oil content and quality.

4. Conclusion

The results of the diagnostic study showed that fertilization is not well controlled by olive growers and olive orchards are not well nourished with essential elements for their production. This was confirmed by soil and leaf analyzes of samples taken from these orchards. Hence need to study the impact of this technique on olive production by conducting field fertilization trials, especially since in Morocco few studies have focused on this aspect.

The results of the olive tree fertilization trials conducted in different regions showed that nitrogen did not improve olive tree productivity parameters in two sites and even negatively affected the olive oil quality. At S2, nitrogen improved yield, yield efficiency, and olive oil content; the best results were obtained with the dose of 0.5 Kg $N.tree^{-1}.year^{-1}$. Phosphorus did not have a significant impact on the olive tree at the three sites. Potassium affected yield and yield efficiency at one site (S3) and had no effect on oil content and quality. An application of 0.5 Kg $K_2O.tree^{-1}.year^{-1}$ allowed a good yield.

The results of three experimentations showed that the effect of mineral fertilization on the olive tree was variable depending on the environment where it is grown (climate and soil of the site), the variety, and the orchard age. The fertilizer input is conditioned by the combination of all these parameters. The results of these field trials remain preliminary given the short duration of experiments. Field trials on olive fertilization should be repeated for several years and in different agro-climatic zones to be able to emerge reliable standards of crop fertilization. However, from this work, it can be concluded that nitrogen and potassium are the two most important elements for olive nutrition and which can affect both its productivity and its quality. The impact of phosphorus on the crop has not been significant, whereas our farmers provide it in high doses compared to the crop's need.

In Morocco, fertilization standards for the olive tree are not yet clearly defined. Research work on this topic seems to be insufficient and should be further developed.

Conflict of interest

The author declares no conflict of interest.

Author details

Karima Bouhafa
National Institute of Agronomic Research, Meknes, Morocco

*Address all correspondence to: karima.bouhafa@inra.ma

References

[1] Martinez Raya A. Suitable land for olive cultivation. Inter. Course on the Ferti. And Intens. Culture Of the Olive, Spain. 1983;**1983**:26-29

[2] Llamas JF. Basis of fertilization in olive cultivation and the olive tree's vegetative cycle and nutritional needs. Int. course on the fertilization and intensive cultivation of the olive. Cordoba. 1983;**1983**:1-25

[3] California Fertilizer Association. 1995. Western Fertilizer Handbook. Eighth Edition

[4] Marin L, Fernández-Escobar R. Optimization of nitrogen fertilization in olive orchards. In: Val J, Montanés L, Monge E, editors. Proceeding of the Third International Symposium on Mineral Nutrition of Deciduous Fruit Trees. Spain: Zaragoza; 1997. pp. 411-414

[5] Tubeileh A, Turkelboom F, Al-Ibrahem A, Thomas R, Tubeileh KS. Modelling the effects of soil conditions on olive productivity in Mediterranean hilly areas. International Journal of Agronomy. 2014;**2014**, Article ID 672123:12

[6] Gargouri K, Mhiri A. Relationship between soil fertility and phosphorus and potassium olive plant nutrition. Options Mediterranean's: Serie A. 2002;**2002**(50):199-204

[7] Tagliavini M, Marangoni B. Major nutritional issues in deciduous fruit orchards of North Italy. Hort Technology. 2002;**12**:26-31

[8] Bouma J. Precision agriculture: Introduction to the spatial and temporal variability of environmental quality. In: Lake JV, Bock GR, Goode JA, editors. Precision Agriculture: Spatial and Temporal Variability of Environmental Quality. Ciba Foundation Symposium, 210. Wageningen, the Netherlands: Wiley; 1997. pp. 5-17

[9] Centeno A, Gómez del Campo M. Response of mature olive trees with adequate leaf nutrient status to additional nitrogen, phosphorus, and potassium fertilization. Acta Horticulturae. 2011;**888**:277-280

[10] Fernandez-Escobar et al. 2009

[11] Garcia CG. Abonado Del olivar. In: Guia Practica de la fertilizacion Racional de los cultivos en Espana. Madrid, Espana: Ministerio de Medio Ambiente y Medio Rural y Marino; 2009. pp. 223-233

[12] Civantos L. La olivicultura en el mundo y en Espana. In: Barranco D, Fernandez-Escobar R, Rallo L, editors. El cultivo del olivo. Mdrid: Mundi-Prensa; 2008. pp. 17-35

[13] Fernandez-Escobar R, Ortiz-Urquiza A, Prado M, Rapoport HF. Nitrogen status influence on olive tree flower quality and ovule longevity. Environmental and Experimental Botany. 2008;**64**(2):113-119

[14] Ferreira J, Garcia-Ortiz A, Frias L, Fernandez A. Los nutrientes N, P, K en la fertilizacion del olivar. X Aniversario Red Cooperativa Europea de Investigacion en Oleicultura. Cordoba; 1984

[15] Fernández-Escobar R, Marin L, Sánchez-Zamora MA, García-Novelo JM, Molina-Soria C, Parra MA. Long-term effects of N fertilization on cropping and growth of olive trees and on N accumulation in soil profile. European Journal of Agronomy. 2009a;**31**:223-232

[16] Rodrigues MA, Pavao F, Lopes JI, Gomes V, Arrobas M, Moutinho-Pereira J, et al. 2011. Olive yields and tree nutritional status during a four year period without nitrogen and boron fertilization. Communications in Soil Science Plant Analysis. 2011;**42**(7):803-814

[17] Jasrotia A, Singh RP, Singh JM, Bhutami VP. Response of olive trees to varying levels of N and K fertilizers. Acta Horticulture. 1999;**474**:337-340

[18] Fernández-Escobar R, Parra MA, Navarro C, Arquero O. Foliar diagnosis as a guide to olive fertilization. Spanish Journal of Agricultural Research. 2009b;7(1):212-223

[19] Cimato A. Effect of agronomic factors on virgin olive oil quality. Olivae. 1990;**31**:20-31

[20] Giménez C, Diaz E, Rosado F, Garcia-Ferrer A, Sanchez M, Parra MA, et al. Characterization of current management practices with high risk of nitrate contamination in agricultural areas of southern Spain. Acta Horticulture. 2001;**563**:73-80

[21] Lopez-Granados F, Jurado-Expositi M, Alamo S, Garcia-Torres L. Leaf nutrient spatial variability and site-specific fertilization maps within olive (Olea europaea L.) orchards. European Journal of Agronomy. 2004;**21**(2004):209-222

[22] Fernández-Escobar R, Beltrán G, Sánchez-Zamora MA, García-Novelo J, Aguilera MP, Uceda M. Olive oil quality decreases with nitrogen over-fertilization. HortScience. 2006;**41**(1):215-219

[23] Therios I. Mineral nutrition of olive trees. Proceedings 2nd International Seminar Recent Advances Olive Industry. 2006:403-410

[24] Righetti TL, Wilder KL, Cummings GA. Plant analysis as an aid in fertilizing orchards. In: Westermann RL, editor. Soil Testing and Plant Analysis. 3rd ed. Book Series no. 3. Madison, WI: SSSA; 1990. pp. 563-601

[25] Fernandez-Escobar R. Las prácticas de la fertilización del olivar en la Cuenca del Mediterráneo. Olivae. 2008;**109**:13-22

[26] Freeman M, Carlson RM. Mineral nutrient availability. In: Sibbett GS, Ferguson L, editors. Olive Production Manual, 3353. 2nd ed. Oakland, CA: University of California Publication; 2005. pp. 75-82

[27] Rodrigues MA, Ferreira IQ, Claro AM, Arrobas M. Fertilizer recommendations for olive based upon nutrients removed in crop and pruning. Scientia Horticulturae. 2012, 2012;**142**:205-211

[28] Fontanazza G. Comment cultiver en vue de la qualité de l'huile. Olivae. 1988;**24**(36):43

[29] Lopez-Villalta LC. Production technique. In: International Olive Oil Council, editor. World Olive Encyclopedia. Sabadell, Spain: EGEDSA; 1996. pp. 145-190

[30] Ben Mimoun M, Gharb M, Ghanem M, Elloumi O. Effects of potassium foliar spray on olive, peach and plum, Part 1: olive experiments. Optimizing Crop Research. 2008;**23**. N°. 17, September, 2008

[31] Arquero O, Barranco D, Benlloch M. Potassium starvation increases stomatal conductance in olive trees. HortScience. 2006;**41**:433-436

[32] Restrepo M, Benlloch M, Fernandez-Escobar R. Influencia del estrés hidrico y del estado nutritivo del olivo en la

absorcion foliar del potasio. In: Direccion General de Invetigacion y Formacion Agraria y Pesquera, editor. Jornadas de Investigacion y transferncia de tecnologia al Sector Oleicola. Cordoba (Spain); 2002. pp. 307-310

[33] Chapin FS. Effects of multiple environmental stresses on nutrient availability and use. In: McConey HA, Winier WE, Pell EJ, editors. Response of Plants to Multiple Stresses. Acad. Press, Inc; 1991

[34] Mengel K. Alternative or complementary role of foliar supply in mineral nutrition. Acta Horticulture. 2002;**594**:33-47

[35] Inglese P, Gullo G, Pace LS. Fruit growth and olive quality in relation to foliar nutrition and time of application. Acta Horticulture. 2002;**586**:507-509

[36] Sarrwy SMA, Mohamed EA, Hassan HSA. Effect of foliar sprays with potassium nitrate and mono-potassium phosphate on leaf mineral contents, fruit set, yield and fruit quality of Picual olive trees grown under sandy soil conditions. American-Eurasian Journal of Agriculture Environment Science. 2010;**8**(4):420-430

[37] Allalout A, and Zarrouk M. Culture hyperintensive de l'olivier dans le monde et applications en Tunisie. HTE N 157-158 Sep/Dec 2013. 2013

Chapter 2

The Olive Tree: A Symbol

Stefano Grego

Abstract

The olive tree is not only the typical plant of the Mediterranean but it is also a tree that constitutes the only culture of global importance. From very old times through the centuries, the traditions of the great oil-growing areas of today were born and consolidated—Greece, Italy, and Spain. The olive tree and oil are an indispensable presence for our daily well-being, as well as a reminder of our oldest and truest history. Olive tree became a symbol of peace and value, an element of strength and purification, of resistance to the ravages of time and wars, the olive tree has always been a transcendent symbol of spirituality and sacredness.

Keywords: multifunctionality, symbol of peace and valor, spirituality and sacredness, significant role in religion

1. Introduction

No other tree or plant, with the exception of wheat, has had the same importance in the history of our species as the olive, capable of influencing nutrition, culture, and religion and contributing to the birth of modern Mediterranean civilization [1]. The unmistakable gnarled and curved branches of olive trees have shaped the Mediterranean landscape and their fruits have provided sustenance to numerous civilizations, from the Middle East to Greece, from Italy to Spain. "Two are the liquids particularly pleasing to the human body: inside the wine, outside the oil," wrote Pliny the Elder in his famous treatise Natural History. "Oil is an absolute necessity and man was not wrong in dedicating his efforts to obtain it."

2. A short history

To understand when the olive tree became essential for the peoples of the Mediterranean, it is necessary to take a journey into the past, going back thousands of years, trying to understand the origins of this special plant. It is believed that the first cultivation of olive trees, as we know them today, began about six to seven thousand years ago in the regions of the Middle East, in an area corresponding to ancient Persia and Mesopotamia. Here, the wild olive tree was domesticated and the oil was produced for the first time. From the Bronze Age, this precious substance spread like wildfire, thanks to merchants, moving between Syria, Lebanon, Israel, Palestine, Egypt, Greece, and Italy, and over time it acquired a growing socio-economic value. Terral et al. [2] analyzed the genetic differences between wild and cultivated olive trees, to trace the origin of the modern European olive tree, the European Olea [3]. According to the scholar, the birth of the modern olive tree is more complex than one might think, deriving as it does from the crossing of 11 different varieties. One thing is certain, however, the olive tree has been part of the Mediterranean flora since time immemorial, fossil pollen of the genus Olea has in fact been found in several Mediterranean countries, such as Macedonia and Greece, while fossilized leaves dating back to about 37 thousand years ago have been found on the Greek island of Santorini (**Figure 1**) [4].

In 2500 BC within the Babylonian code of Hammurabi, the production and trade of olive oil were regulated, while around 1300 BC olive branches embellished the tombs of the pharaohs.

Thanks to the Greeks, the olive tree spread around the Mediterranean. It was considered very important in Greek culture and is mentioned in Greek mythology, "Poseidon, god of the sea, and Athena, daughter of Zeus, goddess of wisdom", competed to win possession of protection over Athens [5]. Poseidon hit the rock with his trident (on which the Acropolis would later rise) and from this, he made a source of seawater come out and a horse faster than the wind. Athena planted the first olive tree, a tree that, for millennia, with its fruits would give a wonderful juice that men could use for the preparation of food, for the care of the body, for the healing of wounds and diseases, and as a source of light for homes (**Figure 2**).

Thucydides (fifth century BC) wrote that "The peoples of the Mediterranean began to emerge from barbarism when they learned to cultivate olive trees and vines."

After the Greeks who favored the spread of the olive tree in the Mediterranean, the Romans took on the task of bringing the cultivation of this plant to greater development. The olive tree was planted everywhere in the empire, so much so that the Roman Empire imposed the payment of taxes in the form of olive oil. Also thanks to them the production process improved, there was a distinction made between different types of oil connected to the moment of pressing, (5 species of oil were identified). Even in Roman mythology, oil had a certain importance in fact it was Hercules who introduced the olive tree from North Africa and then the goddess Minerva would have taught men the art of cultivation and extraction of oil. As usual, with the fall of the Roman Empire, olive cultivation also experienced a period of decline,

The olive tree played an important role in various fields, including in sport—Greek athletes anointed themselves with olive oil before fighting and the winners of the Olympics were crowned with sacred olive branches and rewarded with

Figure 1.
Olive Grove in Santorini.

DOI: http://dx.doi.org/10.5772/intechopen.102827

Figure 2.
Poseidon and Athena.

Figure 3.
Symbolic wreath of olive branches.

richly decorated oil ampoules. Both the Greeks and the Romans used olive oil in the preparation of numerous recipes; moreover, it was widely used for therapeutic, medicinal, balsamic, and detergent purposes or as a fuel for votive lamps (**Figure 3**).

With the advent of the dominion of Rome, the olive tree had one of its greatest moments of glory and extended its range to reach areas in which until that moment, also due to the unfavorable climate, its cultivation had been marginal or nonexistent. The context in which the Berber populations of northern Africa learned the art of grafting olive trees is not known, but we know from Latin sources that they already practiced it when the Romans conquered their lands.

Finally, in the New World, olive trees were introduced soon after their discovery at the end of the fifteenth century. The first olive trees arrived in the Antilles from the port of Seville after the discovery of the New World and since 1560 olive groves have been observed in Mexico and later also in Peru, California, Chile and Argentina [6].

3. The olive and the monotheistic religions

The presence of the olive tree as a symbol and in myths goes back a long way and has its roots in prehistory. Over time, these plants have lent themselves to numerous interpretations—for Homer, the olive tree was a symbol of peace and life. The Greek poet also included it in the Odyssey—it was in fact an olive tree trunk, a plant sacred to Athena, the one with which Ulysses blinded the Cyclops Polyphemus. It was also considered an emblem of strength and victory, in ancient Greece, the winners of the Olympics were offered a wreath of olive trees and a cruet of oil. The deep bond between the Hellenic country and the olive tree is certified by the legend according to which the goddess Athena struck the rock with her spear, giving birth to the first olive tree in the world. Even in ancient Rome, olive oil, an indispensable product in daily life and an ingredient of Roman cuisine, and the plant from which the precious fruits were born were revered. The myth has it that Romulus and Remus, the twin protagonists of the Roman mythological tradition, were born right under an olive tree. The ancient Egyptians believed instead that it was the goddess Isis who revealed the properties of the olive tree to man and taught him the art of cultivating and producing oil.

The Olive tree is so strongly associated with peace, rebirth, and victory, that it is hardly surprising that it has become one of the main symbols of Easter, a sign of rebirth and regeneration [7].

According to the scriptures, it was after the Great Flood that Noah received an olive branch from a dove, which from that moment became the symbol of the promise of rebirth, of a new beginning, and of regeneration [8].

The history of the Olive tree is lost, intertwined, in the history of humanity. All we can get from the symbolism of this tree are some Greek/Roman customs and practices that attest to its importance also for polytheistic religions.

For example, in Greece, the olive tree was considered a sacred tree, to the point that anyone caught damaging or cutting it was exiled. At the Olympics, an olive crown was given to the winner, along with an oil cruet. But not only—in Athens, a particular olive tree was recognized as the first olive tree in the world, which was, therefore, treated with great respect and considered sacred, as connected to the goddess Athena, Patroness of the city.

The custom of crowning with an olive tree was not lost and it seems was also imported to Rome, where this particular crown was used to honor the most valiant citizens, against the more famous laurel wreath which was instead taken as a symbol of victory and domain.

More likely, however, the Catholic symbolism of the Olive tree derives from a synthesis of these pagan meanings and the more well-known Hebrew meaning, mentioned earlier in reference to the Old Testament [9].

It is not clear whether the Olive Tree has any connection with the Tree of the Knowledge of Good and Evil, as when Adam – dying – sent his son Seth to bring him the three seeds of the Tree of Sin, he obeyed and, after his father's death, he planted the seeds on Adam's tomb, from which a cypress, a cedar, and an olive tree grew.

Whatever the origin of the symbol behind this particular plant, it teaches us the need for renewal and regeneration, as well as the great ability to pacify one's feelings with those of God or the Universe, however you want to call it (**Figure 4**).

The only certain thing about the history of the olive tree is that it could not be a more suitable plant in the vegetable kingdom to represent the Christian Easter and the whole spring period in general.

The olive tree and olive oil have always seemed to instill in the observer a profound sense of the sacred. It is no coincidence that they play an important role in the three monotheistic religions of the Mediterranean, Judaism, Christianity, and Islam. The oldest written testimony on the olive tree is found in the Old Testament, the sacred

Figure 4.
Peace dove.

book of Judaism, in the episode of the Great Flood: Noah waited another seven days and again sent the dove out of Ark. In the evening, the dove returned to him: and behold, it had an olive branch in its beak. In Christianity, we remember Palm Sunday, which commemorates the entry of Jesus into Jerusalem welcomed by the crowd waving olive branches and palm trees. The oil is also used in baptism, in the consecration of priests, in the anointing of the sick, and in confirmation. As for Islam—in Paradise, there are two taboo trees, the olive tree and the fig tree.

The very name of Christ means "anointed," a Greek translation of the Hebrew term mašíakh used as an adjective that designates the person who was anointed with oil in the consecration ceremony. In the Christian religion, the olive tree has a strong symbolic value, the olive branch pinched in the dove's beak, for example, marks the end of the flood, symbolizing peace and regeneration. Even in Jewish religious rites, oil is very important and the theme of anointing as a consecration to the Lord returns. A series of food taboos have also emphasized its importance, the Torah, the fulcrum of the Jewish religious tradition, in fact, prohibits the consumption of most animal fats, helping to make olive oil the main ingredient of Jewish cuisine. The olive tree, in addition to being present in Greek mythology in an important way, is also known in the Arab tradition, which, however, based most of its current doctrine on the teachings of Muhammad and also traces to a large extent pre-existing beliefs that, so as happened within the Christian religion regarding pagan cults, they had to be implemented by the Prophet in the global order of the new religion. In the Koran, the sacred text of Islam, the olive tree is defined as "the blessed tree," while the anthropologist Edvard Westermarck, in the essay Ritual and Beliefs in Morocco, wrote: "in Islam, it is the cosmic tree par excellence, center and pillar of the world, symbolizes the universal man, the prophet."

Generally, in religious language, the olive tree has always represented the promised land; it has always been a message of fertility and divine blessing (a message which, among other things, is common to several religions). In old times, olive growing was in fact among the main agricultural activities, and oil, obtained from this tree, along with cereals and wine, represented one of Israel's heritage. The olive tree and the resulting oil were in fact signs of abundance and well-being.

In the Bible, more than in other religious texts, it is present in numerous episodes and has a much stronger symbolism than in other religions. However, the fact that the olive tree is present in the most ancient texts and that it is part of several different and distant cultures historically and geographically, makes us understand how important and ancient the culture is linked to this tree.

4. The olive tree in art

The culture of the olive tree and its fruits has deep ties not only in the gastronomic traditions of the different populations of the countries that overlook the sea "nostrum" but so permeates the civil and religious culture of the various nations that often they have based their own economic survival on the production of olives. Perhaps no tree like the olive has moved from cultivation to culture, each becoming an integral part of the other.

In the course of history and literature, numerous artists and writers have dedicated prose passages and poems to the olive tree, a plant that, due to its strength and structure, has always fascinated man and has become the tree par excellence, the protagonist of stories, tales, and myths.

Ode, canto, what a beautiful image comes to mind when quoting these two words. What if an ode was dedicated to food? It was Pablo Neruda, pseudonym of Ricardo Eliécer Neftalí Reyes Basoalto (1904–1973), poet, diplomat, Chilean politician, Nobel Prize for literature, who generated this great idea. In the Ode to Wine and Other Elementary Odes, Neruda celebrates wine, bread, onion, tomato, oil, potato, and other foods, apparently so mundane, of which we do not realize their immense value. The poet approaches poetry with small things, giving them a new identity.

Evergreen and millennial trees are symbols of peace, hope, and resistance, the olive tree is already present in the sacred texts of the three major religions in the world, and from then on many writers have used this plant as a subject or metaphor in their texts or in their paintings.

With its leaves with iridescent, silvery reflections, whose delicate lines are highlighted by the branches and gnarled trunk, the olive tree represents in the collective imagination a sort of sculpture made by nature itself, also for this reason, in the history of art it has been the privileged subject, a constant source of inspiration for artists, also as a symbolic representation of virtue, dogmas, images, feelings, affections, and emotions [10, 11].

The olive tree has been represented in many religious paintings. The image of the olive tree is often the element with which the artist guides the observer on a path of didactic references

In the historical-artistic context and in particular, in the iconography of the Virgin and the Passion of Christ, the image of the olive tree is often the element with which the artist guides the observer in a path of didactic references, as on the figure of Maria Regina Pacis, parent of the Savior of the world.

An example is Giotto's *Entry of Christ into Jerusalem*, in the Scrovegni Chapel in Padua. The scene is composed of an amiable realism, very evident in the figure of the donkey, placed in the foreground, and in the atmosphere that the image itself, as the artist conceived it, generates in the viewers (**Figure 5**).

Figure 5.
Entry of Christ into Jerusalem *by Giotto. Scrovegni Chapel, Padua.*

In the twentieth century, the olive tree was defined, in all respects, as a natural element of the landscape, as a symbol of a pure, and authentic beauty that everyone, poets, writers, artists, can capture and be aware of its fascinating artistic significance.

Another notable example is the *Annunciazione* by Simone Martini and Lippo Memmi in 1333 (Uffizi Gallery in Florence).

The young archangel, kneeling in the manner of a noble knight, hands the Virgin an olive branch, a symbol of the peace and universal harmony that the unborn child would spread on earth. She wears an elegant damask dress (whose golden color reflects Gabriel's nickname, known as the "messenger of light") and a lively checkered cape (**Figure 6**).

Sandro Botticelli, in the painting *The return of Judith to Bethulia* (c. 1472, Uffizi Gallery in Florence), places an olive branch in Judith's hand, symbolizing the rediscovered peace after the death of the Assyrian king Holofernes. This work, together with the other protagonist of the diptych, *Discovery of the corpse of Holofernes*, constitutes one of the first narrative paintings we know of by Botticelli (**Figure 7**).

The two protagonists are portrayed by Botticelli while they are on the run, with the enemy's head covered with a sheet, and Giuditta, while with one hand she holds the murder weapon or the saber, and with the other, she holds an olive branch, a typical symbol of peace.

Another religious example in which the Passion of Christ and the olive tree are intertwined, just like one of its branches, is that of El Greco's 1590 painting Christ in Gethsemane (National Gallery of London). In his unmistakable style with elongated objects and figures, the plant is a characterizing element as it is placed in the foreground to the left of Christ, thus making the viewer participate in the story. In it, moreover, the olive tree foreshadows the death of Christ and the peace to which humanity is destined by the will of God following the sacrifice of the son.

A further tribute to the sacredness of this tree and its fruits can be found in the marvelous *Madonna dell'olivo* by the Genoese Niccolò Barabino of 1888 in which the very pure Virgin in white hugs the baby Jesus who in turn holds a twig of olive tree, once again a symbol of peace and salvation (**Figure 8**).

Figure 6.
Simone martini annunciazione.

Figure 7.
The return of Judith to Bethulia. Sandro Botticelli, Uffizi Gallery, Florence

The olive tree whose branches take on the same color as the background is placed at the back of the scene and in turn, completes the triad of the embrace (Madonna, Jesus, olive tree) in a global way, thus framing the picture and letting you imagine its continuation. beyond the perimeter.

Still in the same century but 50 years earlier, the landscape current of Realism and Naturalism spreads in France, connected to the school of Barbizon, a small town near the forest of Fontainebleau, whose painters, despite differing in style and temperament, are strongly linked because they share the same desire to discover the

Figure 8.
Madonna dell'olivo *by the Genoese Niccolò Barabino. Church of S. Maria della Cella in Sampierdarena – Genova.*

beauty of Nature. This will be the backbone of a new movement that will soon spread, Impressionism, through which painters bring to completion "[...] the accentuation of the perceptual moment over the fantastic" by painting en plain air.

Based on these assumptions, Van Gogh, the movement's leading exponent, in 1889 dedicated approximately 18 canvases to the representation of the olive tree in the autumn during which he was hospitalized in the psychiatric hospital of Saint-Rémy-de-Provence for serious emotional and nervous difficulties. For him, the olive trees represent life and its cycle, the divine, and how the relationships between man and nature can connect the former with the divine. Furthermore, for the painter, being in harmony with nature means creating moments of idyll and contemplation (**Figure 9**).

National Gallery of Art in Washington summarizes this series:

"In the olive trees – in the expressive power of the ancient and gnarled forms – Van Gogh found the manifestation of the spiritual force that he believed to reside in all of

Figure 9.
Representations of olive trees by Vincent Van Gogh.

nature and his brushstrokes make the ground and the sky alive with the same movement of the rustling leaves, mixed to a shimmer by the Mediterranean wind. The energy in the continuous rhythm communicates to us, in an almost physical way, the living strength that Van Gogh found among the olive trees; that spiritual force that he believed took shape there."

5. A second product: landscape

In reality, the "landscape" is a vast and difficult subject to circumscribe also because it is a concept that has undergone a profound evolution over time. The European Landscape Convention, Florence, 20.X.2000 defined that "The member states of the Council of Europe signatory hereto,.... Noting that the landscape has an important public interest role in the cultural, ecological, environmental, and social fields, and constitutes a resource favorable to economic activity and whose protection, management, and planning can contribute to job creation; Aware that the landscape contributes to the formation of local cultures and that it is a basic component of the European natural and cultural heritage, contributing to human well-being and consolidation of the European identity;...." [12].

The landscape is, therefore, the heterogeneous set of all the elements, processes, and interactions that make up the ecosphere, considered in its unitary and differentiated structure and in which the activities of nature and man are integrated, in their historical dimension. Material, cultural, and spiritual.

"Cultural Route of the Council of Europe" certified in 2005 that "The presence of the olive tree has marked not only the landscape but also the everyday lives of the Mediterranean peoples. As a mythical and sacred tree, it is associated with their rites and customs and has influenced their lifestyles, creating a specific ancient civilization, the "Olive Tree Civilization": the Routes of the Olive Tree follow in the footsteps of this civilization, from Greece toward the Euro-Mediterranean countries. The olive tree dates back millions of years. Wild olive trees, ancestors of the domesticated ones, can still be seen in the Peloponnese, Crete, North Africa, and the Middle East, their places of origin. The relationship between this tree and human civilization has produced an immensely rich, living cultural heritage, embedded in the everyday habits of the Mediterranean people. From gastronomy, with the crucial influence of olive oil, to art and traditions, the social development of these areas has been largely shaped by the olive tree" [13].

The olive tree participates in the formation of a multiplicity of landscapes in relation to the different cultural structures that have been defined in the long process of adaptation of the species to the different environmental characteristics of the places. uneven orography of the most disadvantaged areas, to the more or less specialized systems of the hilly areas, to the intensive monoculture of the plains. In addition to the landscape diversity dictated by the cultivation practices, strongly changing aesthetic connotations are impressed on the territory by the different olive models adopted, the result of that continuous centuries-old adaptation of cultivation techniques to environmental conditions, whether linked to the company structure and to the edaphic and climatic conditions or the economic and social structure that has gone through human history, where the olive trees were associated with a myriad of other crops, which ensured the livelihood of peasant families, reinforcing the typical geometries of poly-cultural systems (olive trees placed at the edge of the vineyards or between the rows, inserted together with almond and carob trees, confined within the vegetable gardens or placed at the edge of citrus groves) or interrupting the monotony of progress of arable land (**Figure 10**).

With the development of olive growing in vast territories of the island, and particularly starting from fifteenth to eighteenth century, the presence of plots with higher specialization, it begins to become more consistent, to the point of characterizing, in subsequent phases, the crop structure of entire territories, especially hilly areas. This singular olive type, defined as "traditional," still constitutes wealth for the area today due to the important role it fulfills in hydro-geological defense and in the qualification of the landscape. As a result of the millennial interaction between different environmental, social, and cultural factors, it is still possible today in many countries to find the numerous systems and landscapes of the olive tree that have accompanied its history.

In the most marginal conditions, on sloping land, on the narrowest terraces, the olive tree participates in the formation of the systems and landscapes of mixed cultivation where these systems survive the rural exodus of farmers.

These olive-growing systems frequently constitute tiles within a mosaic formed by very fragmented agricultural systems of different types and with high landscape diversity. Olive groves which, thanks to the capacity of self-regeneration, typical of the species, have resisted over time giving rise to specimens of large and very large dimensions defined by Pirandello Saracens for their almost legendary antiquity. Where the orographic conditions are more favorable to productive rationalization processes, the olive cultivation becomes more specialized while only partially maintaining the characteristics of traditional olive-growing systems. Some modifications of the cultivation model which concern the containment of the volume and height of the canopy contribute to differentiate it, to favor and economize the practices of defense, pruning and harvesting, and a reduction of the planting distances, which become regular, to increase the productivity of olive groves. It is amongst the olives grown for oil that it is difficult to indicate a single type of plant. As a result of the different densities adopted, the distances and the planting width vary, with evident reflections on the geometry of the olive groves. In many cases, the plant model and its landscape impact depend on the genotype and, in particular, on the bearing of the plants, on their vigor as well as on morphological characteristics, such as shape, size, and color itself, of the leaves.

Literature also highlights the qualities and prerogatives of the olive tree: its resistance to time and bad weather, the usefulness of its fruit, the sense of peace and serenity that it gives to men with the soft and pale green of its branches. The olive tree is a resistant plant—it sinks its roots firmly into the stony ground, it takes many years to become a plant that bears fruit, but, as it was slow in growing, it lasts a long time.

Figure 10.
Some examples of landscapes characterized by the presence of olive trees.

It binds several generations together, giving fathers the satisfaction of having planted a tree not for their own good, but for that of their children and posterity. The fruit of the olive tree is precious because the oil not only gives nourishment but also light in the lamp; and like light, it accompanies us to the deathbed. Undoubtedly the olive tree is a plant that inspires feelings very close to those of the poet: peace, which is the opposite of the hatred depicted in old castles (manors); generosity toward others (the precious gift of its fruit); goodness, which is the opposite of badness.

6. Final considerations

Traditional olive-growing agroecosystems constitute tiles within a mosaic made up of agrarian and semi-natural systems of different types, very fragmented and with high landscape diversity. Even at the farm level, biological diversity remains high both in the case that the olive tree is part of a poly-cultural system and in the case of olive groves conducted in conditions close to semi-natural. Olive trees are grown according to the knowledge and practices that include the use of different types of terracing, cultivation techniques, and genetic varieties that have been maintained for centuries by local communities. This extraordinary landscape made up of olive trees was shaped by the ancient interaction of farmers with the environment.

The countries of North Africa have their own landscape defined by the olive tree. A typical example is the desert area of Tunisia and Morocco where the olive tree naturally grows mainly in the valleys using the little humidity present. Tunisia is a land of olive trees, a place where the olive tree over the millennia has been infused with the culture, economy, cuisine, habits, rhythms, seasons of the nation. Some Tunisians even anoint babies with olive oil (**Figure 11**).

Multifunctionality has now become the strategic choice undertaken by many farms which, at various levels, carry out various activities to respond to the negative effects deriving from a system mainly oriented to the production of material goods of industrial origin. For agricultural enterprises, multifunctionality represents a "new" way of organizing production factors (internal resources) and interacting with external resources (the territory), aimed at pursuing economic, environmental, and social objectives in the medium and long term. Seen from a more general perspective, multifunctionality represents one of the key points in the development process of the agricultural sector and the rural world. The role of agriculture, in fact, for several

Figure 11.
Olive Tree in Tunisian landscape.

years now is no longer exclusively attributable to its function of producing basic necessities but expands through the recognition and performance of other environmental, social, landscape, historical, cultural, etc.

Multifunctionality actually places agriculture, in its renewed value as a producer not only of traditional goods but also of other goods, at the center of the interest of the economy and citizens. Among other things, multifunctionality is not an exclusive trait of small businesses nor, much less, of marginal agriculture, although multifunctionality can be a strategy aimed at improving the remuneration of the small family business. In the countryside and the primary sector, new economic models are being developed that, looking at the past, at peasant values, at traditional resources and methods, are innovating, revisiting schemes, creating new perspectives, including economic ones; it is agriculture in which future and tradition merge and are declined in retro-innovation—drawing on the experience of the past and enhancing previous knowledge, reinterpreting and using them in contemporary contexts and circumstances, to try to give answers to the needs of the present and above all to ensure that they do not turn into the emergencies of the future.

Author details

Stefano Grego
Interuniversity Centre for Research on Sustainable Development (CIRPS), Rome, Italy

*Address all correspondence to: gregostefano2@gmail.com

References

[1] Schicchi R, Speciale C, Amato F, Bazan G, Di Noto G, Marino P, et al. The monumental olive trees as bio cultural heritage of mediterranean landscapes: The case study of sicily. Sustainability. 2021;**13**(12):6767. DOI: 10.3390/su13126767

[2] Jean-Frédéric T, Breton CM, Newton C, Ivorra S. In: Barbieri G, Ciacci A, Zifferero, editors. Les apports décisifs de la morphométrie (éco-anatomie et morphométrie géométrique) et de la génétique (marqueurs moléculaires microsatellites) dans la reconstruction de l'histoire de le culture et de la domestication de l'olivier. Le origini dell'olivicoltura in Toscana: nuovi percorsi di ricerca tra archeologia, botanica e biologia molecolare. In book: Eleiva, Oleum, Olio. DonChisciotte; 2010

[3] Breton CM, Medail F, Pinatel C, Berville A. De l'olivier a l'oléastre: Origine et domestication de l'Olea europaea L. dans le Bassin méditerranéen. Cahiers Agricultures. 2006;**15**(4):329-336

[4] Breton CM, Warnock P, Berville AJ. Origin and history of the olive. In: Olive Germplasm - The Olive Cultivation, Table Olive and Olive Oil Industry in Italy. IntechOpen; 2012. DOI: 10.5772/51933

[5] Morford MPO, Lenardon RJ, Sham M. Classical Mythology, International Edition. 10th Edition. Oxford University Press, Academic. 9780199997398 Paperback 07 August 2014

[6] Torres M, Pierantozzi P, Searles P, Rousseaux MC, García-Inza G, Miserere A, et al. Olive cultivation in the southern hemisphere: Flowering, water requirements and oil quality responses to new crop environments. Frontiers in Plant Science. 2017;**8**:1830

[7] Moselle Bryan R. The symbolic and theological significance of the olive tree in the ancient near east and in the Hebrew scriptures. PhD Thesis. Pretoria: University of Pretoria; 2016

[8] Lupis J. The Olive Tree. Xulon Press; 2014. Paperback. ISBN: 1629521620

[9] Frank CE. Grafted onto the olive tree: The Biblical Basis for a Cross-Religious Mission Strategy. Create Space Independent Publishing Platform; 2013. p. 106. Christian book

[10] Gixhari B. Olive in history and art. In: International Conference 'The Adriatic Olive Grove: Risk Prevention, Sustainability, Learning'. Vlora, Albania; 2014. DOI: 10.13140/2.1.1760.2885. ISBN-13: 978-1482635188

[11] Ebrahimi N, Motavali H, Branch G. Olive, the symbol of resistance in contemporary Arabic poetry. International Journal of Humanities and Cultural Studies. Special Issue 1 February 2016:880-890. ISSN: 2356-5926

[12] Council of Europe. Landscape and sustainable development: Challenges of the European landscape convention; 2006

[13] Available from: https://www.coe.int/en/web/cultural-routes/the-routes-of-the-olive-tree 2005

Chapter 3

Optimization of the Olive Production Chain through Optical Techniques and Development of New Cost-Effective Optical Systems Inspired by Agriculture 4.0

Valentina Giovenzana, Alessia Pampuri, Alessio Tugnolo, Andrea Casson, Riccardo Guidetti and Roberto Beghi

Abstract

Industry 4.0 is characterized by autonomous decision-making processes, monitoring assets and processes in real time and to real-time connected networks through early involvement of stakeholders. In this scenario, there is a growing interest and a need of innovation also in the agri-food system in the production processes and quality control through the development of new interconnected sensors (IoT approach). Hardware minimization, as well as software minimization and ease of integration, is essential to obtain feasible robotic systems. A substantial change in measurement methodologies is therefore ongoing, and it is of interest the opportunity to replace the consolidated analytical techniques, based on laboratory analyses, with methods based mainly on physical approaches of rapid execution, of limited invasiveness, and with high environmental sustainability. These approaches should be applicable directly in the field or in operative environment, allowing the creation of big databases characterizing the samples, particularly large and shared through the data cloud. This chapter will aim to overview the theoretical principles of the most important technologies applied to the olive oil sector presenting some case studies and will be focused on the future perspective for all operators of the olive sector who want to use a sustainable approach and olive-growing 4.0.

Keywords: agriculture 4.0, optical analysis, Vis/NIR spectroscopy, chemometrics, sensors, qualitative parameters, green technology, machine learning, simplified system

1. Introduction

The agri-food system is increasingly showing the need to innovate production processes and the related quality controls through the use of new technologies and the use of innovative sensors that could be interconnected, approaching what is called industry 4.0. In this context, and in particular in agriculture 4.0, emerging technologies such as artificial intelligence, big data, Internet of Things (IoT) are presented as a solution to the new challenges associated with food production. It is a digitization of all agricultural systems capable of increasing yields by reducing inputs and labor requirements. Furthermore, these technologies are capable of improving the health of the environment by enabling the production of a higher amount of food on the existing land while saving further land conversions and increasing eco-efficiency [1].

Obtaining high-quality and safe agricultural and food products is now an essential condition for both producers and consumers who are more involved and interested in the various aspects concerning food production. Therefore, the agri-food industry is currently concentrating on the production of healthy products that at the same time meet the market demands, and to do this it is essential to carry out punctual and precise quality controls on the products [2].

The analytical methods currently available to assess quality require time and above all are destructive techniques or laboratory chemical analyses that also involve the use of reagents. Nondestructive techniques based on optical properties and visual evaluations of food matrices are now being used all over the world as a response to these needs.

One of the most widespread techniques is undoubtedly visible and near-infrared (vis/NIR) spectroscopy, which is based on the measurement of the variation in the spectral characteristics of a sample irradiated with electromagnetic radiation in the visible and in the near-infrared range (400–2500 nm). The variations of the spectral characteristics in a matrix can be recorded in different modalities according to the characteristics of the product but also according to the characteristics of the instruments used. Spectroscopy for analyzing agricultural and food products has proven to be an exceptional and rapid tool with little or no sample preparation [3].

This type of nondestructive technique guarantees the reduction and, in some cases, even the elimination of the use of solvents, which are instead necessary to carry out traditional chemical laboratory analyzes. Compared with vis/NIR technology, chemical techniques require a lot of time, sample preparation, and the use of chemical reagents influencing both the cost aspect and the environmental impact aspects. Moreover, in recent years, research tends to pay attention also to on/in/at-line applications, and vis/NIR spectroscopy offers several opportunities for quality control during processes: the replacement of the analytical tools and reagents related to chemical analyses with one vis/NIR spectrometer could reduce the environmental impact of analyses [4].

Vis/NIR spectroscopy is just one example of the numerous techniques that are being implemented in these fields. Paragraph 2 of the chapter will analyze the principles of the most common nondestructive techniques used in the agri-food industry, paragraph 3 will focus on the applications of these techniques in the optimization of the olive production process, and finally, fourth paragraph will illustrate the portable prototypes and future prospects of simplified optical devices.

2. Main optical nondestructive approaches and data analysis

2.1 Vis/NIR and NIR spectroscopy

Among the nondestructive techniques, spectroscopic analyses in the visible–near infrared (vis/NIR) and near infrared (NIR) regions are widely used in different fields.

Since the early 1970s, various instruments have been built that are able to exploit these technologies: instruments that acquire the sample spectrum in a specific wavelength range and record the average spectrum of a single defined area of a sample.

Vis/NIR and NIR spectroscopies are used to acquire punctual information on the nature of the functional groups present in a molecule by exploiting the interaction between light and the structure of a sample. The electromagnetic radiation is in fact able to promote vibrational transitions in the molecules. Spectra in the visible region (between 400 and 700 nm) and spectra in the near infrared region (between 700 and 2500 nm) are composed of combination and overtone bands related to absorption frequencies in the mid-infrared region (MIR, between 2500 and 50,000 nm).

All these combinations and overtone bands correspond to the frequencies of the vibrations between the bonds of the atoms that compose the molecules of the analyzed matrix. Each matrix or material is a unique seal of atoms so there are no two compounds capable of producing the same vis/NIR spectra. Through the use of chemometric statistical analyses, it is possible to use spectroscopy as an excellent tool to perform quantitative analyses. A peculiar aspect of this technique is that it does not require sample preparation, thus offering a valid alternative to traditional chemical or physical analytical methods, which instead requires time and the use of solvents or other materials. The data deriving from the spectroscopic analyses are complex and require specific statistical analyses to obtain the information of interest [5].

2.1.1 Principles and instrumentation

The chemical composition and physical characteristics of a sample determine reflection, absorption, or transmission of the electromagnetic radiation. The reflected light could cause specular reflection shine (to be avoided), while diffuse reflection is produced by rough surfaces. These reflection phenomena provide information on the sample surface. More interesting could be the scattering resulting from multiple refractions within the material. The sample heterogeneity is highly influencing the scattering effects. Also, size, shape, and microstructure of the particles have an effect on scattering.

Scattering affects the reflected spectrum, while the sample shape is more related to the absorption process. The bands of absorption in the NIR region are mainly overtones and combination bands of the fundamental absorption bands in the IR region, deriving from vibrational and/or rotational transitions. In the case of complex matrices such as foods, multiple bands and the effect of the widening of the peaks determine vis/NIR and NIR spectra with a wide coverage and few acute peaks.

To acquire a spectrum, it is necessary to use an instrument called a spectrophotometer, which consists of a light source, an accessory to present the sample, a monochromator, a detector, and optical components. Spectrophotometers are classified according to the type of monochromator: it is a device able to decompose a single polychromatic light beam into several monochromatic light beams (that contains waves of a single frequency), thus allowing to analyze the intensity as a function of wavelength.

In a filter instrument, the monochromator is a wheel holding absorption or interference filters and has a limited spectral resolution. In a scanning monochromator instrument, a grating or a prism is used to separate the individual frequencies of the radiation entering or leaving the sample so the radiation at the different wavelengths can hit the detector.

Spectrophotometers based on Fourier transform use an interferometer to generate a modulated light beam. Using the Fourier transform, the light reflected or transmitted by the sample is converted into a spectrum. The most diffused systems use the Michelson interferometer, but also polarization interferometers are employed in the optical bench of some instruments. The photodiode array (PDA) spectrophotometers have a wide diffusion; these systems are based on a fixed grating, which focuses the radiation onto a silicon array of photodiode detectors. The systems based on laser do not use monochromator but different laser sources or a tunable laser. Finally, acoustic optic tunable filter (AOTF) and liquid crystal tunable filter (LCTF) instruments are available on the market. AOTF uses a diffraction-based optical-band-pass filter easily tunable varying the frequency of an acoustic wave propagating through an anisotropic crystal medium. LCTF instruments use a filter to create interference in phase between the ordinary and extraordinary light rays passing through a liquid crystal. The combination of different tunable stages in series can result in a high resolution.

2.2 Computer vision and image analysis

One of the limitations of spectroscopic analyses is the punctual measurement and therefore the inability to provide information on the distribution of an object. Depending on the uniformity of the qualitative attribute to measure, it may be necessary to repeat the spectral acquisition in several points of the sample.

In order to get the spatial distribution, vision technique is a solution. With the huge development of imaging technology, computer vision results attracting for agri-food industry. A large number of applications have been developed for quality inspection, classification, and evaluation of agri-food products [6, 7]. Image data can reflect many external features of a sample such as color, shape, size, surface defects, or contaminations. Computer vision has been applied to solve various food engineering problems ranging from quality evaluation of foodstuffs to quality attributes unavailable to human evaluators.

Computer vision tools are powerful but not much useful for in-depth investigation of internal characteristics. This is due to the very limited capability to provide spectral information with this technique.

2.2.1 Multispectral and hyperspectral images

RGB images, represented by three overlapping monochrome images, are the simplest example of multichannel images. The multispectral images are usually acquired in three/ten spectral bands including in the range of visible, but also in the range of infrared, fairly spaced. In this way it is possible to extract a larger amount of information from the images respect to those normally obtained from the RGB image analysis. The bands that are used in this analysis are the band of blue (430–490 nm), the band of green (491–560 nm), the band of red (620–700 nm), and the band of NIR and MIR. Different spectral combinations can be used depending on the research aims. The combination of NIR-R-G (near infrared, red, green) is often used to identify green areas, for example, from satellite images. On the contrary, the combined use of NIR-R-B (near infrared, red, blue) is very useful to analyze fruit ripeness, thanks to chlorophyll absorption in the red range. Finally, the combination of NIR-MIR-blue (NIR, MIR, and blue) could be used to observe the sea and ocean depth.

Hyperspectral imaging (HSI) is a powerful tool combining spectroscopy and imaging into a three-dimensional data structure (hypercube). The HSI is based on the acquisition of a large number of images at different spectral bands, allowing analysis of each pixel obtaining at the same time a spectrum associated with it. The data structure of a hyperspectral image is data cube, considering two spatial directions and one spectral dimension.

Hyperspectral technology can integrate the advantages of conventional digital imaging and spectroscopy to obtain both spatial and spectral information from an object simultaneously.

In recent years, HSI has been applied to food safety and quality detection, because the technology can achieve rapid and nondestructive detection of food, and the requirement to experimental condition is low [8].

HSI has opened up new possibilities within agri-food analysis, in particular Liu et al. [9] outlined detailed applications in various food processes including cooking, drying, chilling, freezing and storage, and salt curing, emphasizing the ability of HSI technique to detect internal and external quality parameters in different food processes [9].

Using HSI, the hypercube can be acquired in reflectance, transmission, and fluorescence. Nevertheless, the most used acquisition techniques for spectral images are reflectance, transmission, and emission, considering the scientific works published. HSI has many advantages, e.g., the huge time savings that can be obtained for the application to industrial production processes. The advantages of HSI for the agri-food sector can be listed as follows: (i) not necessary sample preparation; (ii) noninvasive methodology that avoids sample losses; (iii) economic value related to time, labor, reagents, savings, and a strong cost-saving for waste treatment; (iv) for each pixel of the sample is acquired the full spectrum and not only few wavelengths; (v) many constituents can be predicted at the same time simultaneously; (vi) special region of interest could be selected and analyzed.

The hypercube generated by using HSI provides a large dataset. The information derived from the hypercube may contain also redundant information. This data abundance may cause a high computational load due also to the long acquisition time. Therefore, it is desirable to reduce this load at acceptable levels, considering the application of HSI for real-time application. For this purpose, the spectral image is appropriately reduced using chemometric data processing, mainly selecting the most informative wavelengths. Using the selected spectral bands, a multispectral system can be envisaged for application at industrial level.

2.3 Chemometrics in agri-food sector

Chemometrics is defined as a branch of chemistry that studies the application of mathematical or statistical methods to chemical data. The International Chemometrics Society (ICS) defines it as a chemical discipline that uses mathematical and statistical methods to: design/select optimal procedures and experiments, provide maximum chemical information by analyzing data, give a graphical representation of this information, in other words, information aspects of chemistry. Chemometrics is essential for processing multivariate data obtained by optical techniques and for obtaining useful information for solving problems related to spectral noise.

One of the most used techniques is the Principal Component Analysis (PCA), also known as the Karhunen-Loève transform. It is an unsupervised exploratory qualitative analysis technique that allows reducing the more or less high number of variables describing a set of data to a smaller number of latent variables, limiting the loss of information.

Other chemometric techniques used extensively in these fields are supervised techniques, techniques that require method validation and that are used to obtain the quantitative prediction of the parameters of interest. Among these we find regression techniques such as Partial Least Square (PLS) regression or Multiple Linear Regression (MLR). The models developed using these techniques must then be tested using independent samples as validation sets to verify the accuracy and robustness of the model.

3. Application of nondestructive techniques for the optimization of the olive production process and enhancement of by-products

Agricultural products are converted into food products by using different processes. The process to achieve the best performance is carried out considering both efficiency and the target quality of the final food product, in order to be competitive on the market. The production of a high-quality extra virgin olive oil (EVOO) could be reached considering an optimization of the different production steps: olive harvesting and handling; milling operation to be done in a short time after harvesting; use of a modern milling plant equipped with suitable technologies to control process conditions. A high level of control of the standard operating conditions is a crucial aspect to avoid process failures and to maintain the highest final product's quality.

During the ripening process, the olives undergo the variation of various physical parameters such as weight, color, pulp-to-stone ratio, and texture and also of chemical parameters such as oil content, fatty acid composition and polyphenol, tocopherols, and sterols content. These characteristics are of great importance because they influence the quality, the yield, and the shelf-life of olive oil and of the by-products of olive production. Olive oils deriving from overripe fruits, for example, have a reduced shelf-life due to the increase in polyunsaturated fatty acids and the decrease in the total content of polyphenols. In particular, in the olive oil extraction chain, process control and management determine the conditions for producing high-quality oil, which is essential both to maintain consumer confidence and to evaluate potential plant yield losses. The flow sheet of the process is based on the following steps: olive cleaning, crushing to obtain a paste, paste malaxation, solid liquid separation, and liquids separation. Solid–liquid separation is a crucial aspect of the entire process. It is based on the separation of the solids (called pomace) from the other components, namely oil and wastewater.

It is important to have online information on the oil content of the olives to set corrective actions during the process in order to reach the best extraction performance. Nowadays the consolidated analysis protocol is based on the Soxhlet method to analyze the oil content in olives, pomace, and pate. This protocol requires a time-consuming drying step, followed by an extraction based on the use of solvent.

For this issue, the Soxhlet method is often substituted in routine analyses by Nuclear Magnetic Resonance (NMR) spectroscopy. Also this procedure is not sufficiently fast due to water interference (the olive pomace must be completely dry). Consequently, this method is unsuitable for an online application.

A precise monitoring of the intermediate products between the olives entering and the oil outlet (the paste, the pomace, and the pate) is crucial for control of the process progress. It is useful to establish correlations among olives, paste, pomace, patè, and oil. For this aim, rapid and possibly easy-to-use technologies are required to assess olive ripening and the characteristics of the by-products. In this way an early detection of possible failures and a continuous monitoring of the production process during its crucial steps result in an adequate control of the oil quality and yield. From this point of view, nondestructive optical applications could greatly help the sector.

Several studies have highlighted the enormous opportunities offered by NIR spectroscopy in terms of applications for quality control during the process, performing on/in/at-line measurements on olive fruits, on pastes, and on oils [10]. Researchers tend to focus attention on the online applications of noninvasive technologies in order to reduce the gap between laboratory scale experimentation and the olive milling industry [11]. A number of studies applying different vibrational techniques in the olive oil chain can be found in the literature, mainly with the aim of standardizing the procedure for an application as official control of the end product [12]. For this purpose, it is crucial to evaluate the optimal spectral range to be used, and the chemometric methods to be performed to obtain robust predictive models for the estimated parameters. On intact olives, Beghi et al. [13] studied the capability of portable vis/NIR and NIR spectrophotometers to investigate different texture indices for the characterization of olive fruits entering the milling process. Salguero-Chaparro et al. [14] used NIR spectroscopy for the online determination of the oil content, moisture, and free acidity performing measurements directly on intact olives.

NIR was used for the analysis of olive by-products (e.g., olive pomace) performing research studies both in lab-scale and in processing mill lines. Barros et al. [15] applied FT-NIR spectrometry (1000–2500 nm) in combination with partial least squares regression for direct, reagent-free determination of fat and moisture content in milled olives and olive pomace; while Allouche et al. [16] used an optical NIR sensor coupled with artificial neural network for online characterization of oil and virgin olive oil to optimize the process. Finally, Giovenzana et al. [17] verified whether vis/NIR spectroscopy could be used to predict the oil content of intact olives entering the mill and of olive paste, pomace, and paté during the milling process.

Multispectral and hyperspectral systems were applied for monitoring the ripening process [18, 19] or on olive oil samples to estimate acidity, moisture, and peroxides by using online system [20] or to discriminate flavored olive oil [21]

4. Portable prototypes and future perspectives toward simplified systems

Having demonstrated the effectiveness of nondestructive analyzes, some problematics remain related to the costs and the dimensions of the instrumentation, two factors that prevent or severely limit some applications of these tools. Research and innovations are allowing these devices to reduce size and weight: devices tend to be more compact and portable. In order to support small producers, systems that are at the same time simple to use and that have a low cost are desirable, so as to make these technologies usable to all and allow real-time evaluations of qualitative and quantitative parameters [22].

Nowadays, chapter authors are working on designing and developing of a simplified LED device for intact olives quality evaluation. A first version of a fully integrated, LED prototype was built and now results patent pending (**Figure 1**).

The peculiar sensory and nutritional characteristics of olive fruits have led to a sharp boost of the demand for the main derivative products in traditional producing areas and elsewhere in the world. Several destructive, expensive, time-consuming, and not sustainable techniques have been used to assess the degree of olives ripeness. To at least partially replace these types of analyses, in 1975, a Maturity Index (MI)

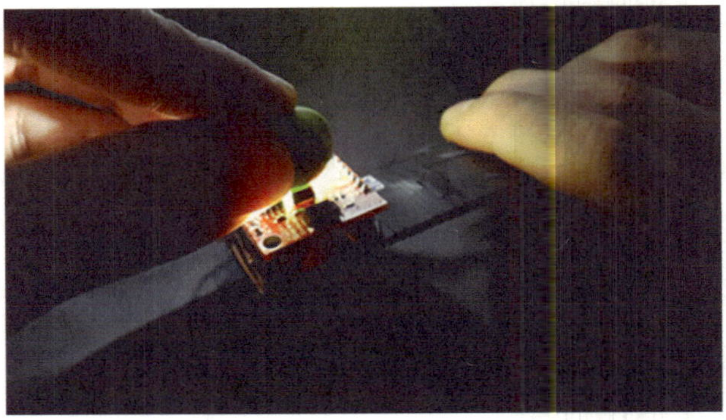

Figure 1.
First version of a simplified LED prototype during optical acquisitions on olives.

was been proposed by Uceda and Frias. This methodology is based on an inexpensive and easy destructive procedure for a visual determination of the best harvesting time. The method is based on color changes of olive skin and flesh; the protocol foresees to classify 100 olives into eight groups, from intense green (category 0) to black with 100% purple flesh (category 7). Despite this protocol being largely used, MI is highly dependent to the operator experience and could be affected by human error. Moreover, olives color changes are very different among cultivars and during the ripeness evolution.

The aim of this research was to design, build, and test cost-effective and user-friendly devices able to optically predict the olive oil and moisture content in olive fruits in order to support small-scale growers in planning the optimal harvest date.

The prototype device is composed of tuned photodiode arrays, interference filters, LEDs, optics and incorporates MEMS (microelectromechanical systems) sensors for spectral measurement in the visible (vis) and short-wave near-infrared (SW-NIR) region.

Therefore, the vision on the application of this sensor can solve several problems in the field of olive growing. Firstly, it can objectify the evaluation of the quality of the olives in the field (to identify the ideal moment of harvesting) and before the milling process to define the correct price of the olives. Secondly, the logistics inside the mill is not easy to be managed. For instance, a preventive evaluation of the maturation parameters could avoid prolonged stop of olives bins in the receiving areas, which causes the deterioration of the product. Finally, the LED prototype could address to olives classification, in terms of qualitative attributes (**Figure 2**), which is useful for high-added-value olive oil productions.

This new generation of optical devices could be a starting point to build a new concept of cost-effective sensors. The stand-alone instrument should be able to acquire and predict the most important ripening parameters directly from measurements in field. This approach could allow olive maturation monitoring bringing the laboratory directly into the field without picking the olive and reducing sampling waste.

The integration of simple multivariate models in the microcontroller software would be easy calculate and visualize the real-time values of the predicted parameters directly on the device to support operators decision-making with objective numbers.

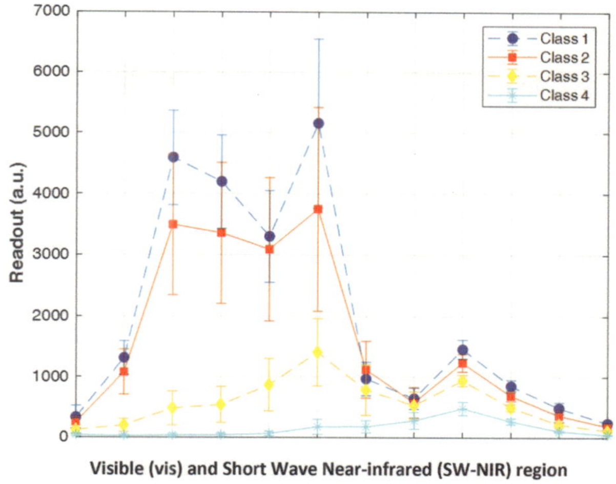

Figure 2.
Average optical readouts and relative standard deviations from each olive ripening class.

5. Conclusions

Among the different available techniques, vis/NIR and NIR spectroscopy and hyperspectral imaging are valid tools for monitoring of qualitative parameters and for maturation control in olive oil sector. The optical instruments currently on the market are mainly laboratory instruments with dimensions and costs that are not suitable for use in real pre- and post-harvest applications, in particular for SME. To overcome this problem, research has concentrated in recent years on feasibility studies and simulations of simplified systems. These studies have been focused on the preliminary design of systems dedicated to single types of product, aiming at a reduced size and low cost.

At the same time, the development and diffusion of cost-effective and increasingly high-performance hardware have opened up new research opportunities envisaging new systems to support optical measurement for the control and management of the pre- and post-harvest processes.

Therefore, further studies both for model improvement and for the design of the system are needed. In a view of olive-growing 4.0, a similar tool based, for example, on a prototype using specific LED for the illumination will lead to quick and accurate analyses in order to get a useful monitoring of the ripening process. In this way it will be possible to estimate the best harvest period and to provide objective features to the operators in terms of quality attributes.

Author details

Valentina Giovenzana*, Alessia Pampuri, Alessio Tugnolo, Andrea Casson,
Riccardo Guidetti and Roberto Beghi
Department of Agricultural and Environmental Sciences—Production, Landscape, Agroenergy, University of Milan, Milan, Italy

*Address all correspondence to: valentina.giovenzana@unimi.it

References

[1] Rose DC, Wheeler R, Winter M, Lobley M, Chivers CA. Agriculture 4.0: Making it work for people, production, and the planet. Land Use Policy. 2021; **100**:104933

[2] El-Mesery HS, Mao H, Abomohra AEF. Applications of non-destructive technologies for agricultural and food products quality inspection. Sensors. 2019;**19**(4):846

[3] Liu D, Zeng XA, Sun DW. Recent developments and applications of hyperspectral imaging for quality evaluation of agricultural products: A review. Critical Reviews in Food Science and Nutrition. 2015;**55**(12):1744-1757

[4] Casson A, Beghi R, Giovenzana V, Fiorindo I, Tugnolo A, Guidetti R. Environmental advantages of visible and near infrared spectroscopy for the prediction of intact olive ripeness. Biosystems Engineering. 2020;**189**:1-10

[5] Beghi R, Buratti S, Giovenzana V, Benedetti S, Guidetti R. Electronic nose and visible-near infrared spectroscopy in fruit and vegetable monitoring. Reviews in Analytical Chemistry. 2017;**36**(4). DOI: 10.1515/revac-2016-0016

[6] Lukinac J, Jukić M, Mastanjević K, Lučan M. Application of computer vision and image analysis method in cheese-quality evaluation: A review. Ukrainian Food Journal. 2018;**7**(2):192-214

[7] Narendra VG, Hareesha KS. Quality inspection and grading of agricultural and food products by computer vision-a review. International Journal of Computers and Applications. 2010;**2**(1):43-65

[8] Zhu M, Huang D, Hu XJ, Tong WH, Han BL, Tian JP, et al. Application of hyperspectral technology in detection of agricultural products and food: A review. Food Science & Nutrition. 2020;**8**(10):5206-5214

[9] Liu Y, Pu H, Sun DW. Hyperspectral imaging technique for evaluating food quality and safety during various processes: A review of recent applications. Trends in Food Science and Technology. 2017;**69**:25-35

[10] Armenta S, Moros J, Garrigues S, Guardia MDL. The use of near-infrared spectrometry in the olive oil industry. Critical Reviews in Food Science and Nutrition. 2010;**50**(6):567-582

[11] Beltran Ortega J, Martinez Gila DM, Aguilera Puerto D, Gamez Garcia J, Gomez OJ. Novel technologies for monitoring the in-line quality of virgin olive oil during manufacturing and storage. Journal of the Science of Food and Agriculture. 2016;**96**(14):4644-4662

[12] Nenadis N, Tsimidou MZ. Perspective of vibrational spectroscopy analytical methods in on-field/official control of olives and virgin olive oil. European Journal of Lipid Science and Technology. 2017;**119**(1):1600148

[13] Beghi R, Giovenzana V, Civelli R, Cini E, Guidetti R. Characterisation of olive fruit for the milling process by using visible/near infrared spectroscopy. Journal of Agricultural Engineering. 2013;**44**(2):e8

[14] Salguero-Chaparro L, Baeten V, Fernández-Pierna JA, Peña-Rodríguez F. Near infrared spectroscopy (NIRS) for on-line determination of quality parameters in intact olives. Food Chemistry. 2013;**139**(1-4):1121-1126

[15] Barros AS, Nunes A, Martins J, Delgadillo I. Determination of oil and water in olive and olive pomace by NIR and multivariate analysis. Sensing and Instrumentation for Food Quality and Safety. 2009;**3**:180-186

[16] Allouche Y, López EF, Maza GB, Márquez AJ. Near infrared spectroscopy and artificial neural network to characterise olive fruit and oil online for process optimisation. Journal of Near Infrared Spectroscopy. 2015;**23**(2):111-121

[17] Giovenzana V, Beghi R, Romaniello R, Tamborrino A, Guidetti R, Leone A. Use of visible and near infrared spectroscopy with a view to on-line evaluation of oil content during olive processing. Biosystems Engineering. 2018;**172**:102-109

[18] González-Cabrera M, Domínguez-Vidal A, Ayora-Cañada MJ. Hyperspectral FTIR imaging of olive fruit for understanding ripening processes. Postharvest Biology and Technology. 2018;**145**:74-82

[19] Martínez Gila DM, Navarro Soto JP, Satorres Martínez S, Gómez Ortega J, Gámez García J. The advantage of multispectral images in fruit quality control for extra virgin olive oil production. Food Analytical Methods. 2022;**15**(1):75-84

[20] Gila DM, Marchal PC, García JG, Ortega JG. On-line system based on hyperspectral information to estimate acidity, moisture and peroxides in olive oil samples. Computers and Electronics in Agriculture. 2015;**116**:1-7

[21] Romaniello R, Baiano A. Discrimination of flavoured olive oil based on hyperspectral imaging. Journal of Food Science and Technology. 2018;**55**(7):2429-2435

[22] Pampuri A, Tugnolo A, Giovenzana V, Casson A, Guidetti R, Beghi R. Design of cost-effective LED based prototypes for the evaluation of grape (Vitis vinifera L.) ripeness. Computers and Electronics in Agriculture. 2021;**189**:106381

Section 2

Olive as a Fruit

Chapter 4

Incorporation of Trigonella Foenum-Graecum Seed Powder in Nitrite-Free Meat Emulsion Systems with Olive Oil: Effects on Color Stability

Theofilos Frangopoulos

Abstract

The replacement of nitrites and starch from Trigonella seed powder in the percentage of myoglobin and metamyoglobin as well as in the color factors (L^*, a^*, b^*) in meat emulsions with olive oil was evaluated. The meat emulsions were prepared on the basis of complete replacement of sodium nitrite ($NaNO_2$) and starch with Trigonella seed powder, where the fat was removed by the Soxhlet method. Thus, two samples emerged, namely, the first sample that was the control and contained 3% starch and sodium nitrite (Starch + $NaNO_2$) in the amount of 150 ppm and the second sample containing Trigonella at 3% (Dtfg) where the fat was removed by the Soxhlet method. The Dtfg sample had a higher percentage of oxymyoglobin ($P < 0.05$) throughout the maintenance period and lower percentages of metamyoglobin ($P < 0.05$) up to the fifth day of maintenance compared to the Starch + $NaNO_2$ sample. The factors L^* (brightness) and a^* (red color) decreased more strongly in the Starch + $NaNO_2$ sample compared to the Starch + $NaNO_2$ sample.

Keywords: meat emulsions, olive oil, Trigonella foenum-graecum, nitrites, green label, color preservation, antioxidants

1. Introduction

The genus Trigonella belongs to the family of cumin (fabaceae or leguminosae), which probably got its name from the triangular shape of the leaves of its flower [1]. More specifically, in the work of Hutchinson [2], it is stated that the genus Trigonella is a member of the subfamily Trifoliae, together with the genera of alfalfa (Medicago), clover (Trifoliae), honeysuckle (Melilotus), and the genus Factorovekya, as the genus Trigonella is subdivided into many known species, of which the species Foenum-Graecum is the best known probably due to its actions in many areas of human activity. Historically, the species Foenum-Graecum appeared around 1500 BC in ancient Egypt from writings available—which happens to be an area of great interest for the medicinal applications of plant species—and mainly, the seeds of the plant were used for therapeutic and embalming purposes [3]. However, the Latin name of this species, Foenum-Graecum, given by the scientists of the time, was attributed to the wider region of ancient Greece. Thus, in the work of Miller [4], it is mentioned that the well-known Greek physician Dioscouridis and "father" of pharmacology from Cilicia, from 65 AD, had included in his dissertation entitled "Materia Medica" the use of the plant for making ointments. The plant trigonella is also mentioned in the medical practice of Hippocrates [3].

It is important to note that technological development and the increase in per capita income from the second half of the twentieth century onward led to the abuse of meat and cold cuts. Specifically, for our country, the per capita consumption of meat products ranges between 8 and 10 kg [5]. The excessive consumption of meat and meat products, in addition to the adverse effects it can have on the body due to chemical additives, leads to a number of negative health effects due to the fat contained in it as well as cold cuts. In particular, Article 91 of the Food and Beverage Code stipulates that the maximum percentage of fat that can be contained in heat-treated meat sausages must not exceed 30% of the product as it stands (with the exception of mortadella, which can reach up to and 35%). The fat contained in these cold cuts, which mostly comes from the back of pigs (lard), plays a very important role in production from both physicochemical and microbiological and organoleptic points of view. However, its high content of saturated fatty acids makes it dangerous for the development of cardiovascular disease and obesity. In this light, scientific research has focused on two directions: one is to reduce the percentage of fat as it is in meat products, and the other is to change the profile of fatty acids by adding fat rich in monounsaturated and polyunsaturated fatty acids. Thus, the addition of olive oil to meat products over the last two decades has been a very interesting scientific and industrial challenge with many approaches in the international literature [6–9].

In recent years, the effects of various food additives on the human body have been studied. In particular, many studies have been conducted, which deal with the negative effects of various chemical additives on the body and the need to find new ingredients of natural origin, which can replace the previous ones leading to the idea of functional food, i.e., the removal from the conceptual characterization of food as food intake and calories necessary for survival, but the consumption of foods that in addition to calories have potentially positive effects on human health in the long and short terms. Moving in this direction, research on the negative effects of chemical additives on the production of meat cold cuts has led to the discovery of harmful compounds created by the intake of nitrites and nitrates reduced to nitrites, which are added to cold cuts. Their association with free amino acids in the body leads to the formation of some very harmful substances with carcinogenic activity known as nitrosamines [10]. Thus, in recent years, a large field of research has been developed around the addition of plant extracts that can replace nitrite in these products [11–13]. Polyphenols are one of the most widespread and numerous groups of bioactive components with a very wide distribution in the plant kingdom and great diversity among different species of plant tissues. They are the products of the secondary metabolism of plants, which means that they are not a primary growth factor for the physiology of the young plant, but they play an important role in the subsequent metabolic and physiological activity of the plant organism. Polyphenols can be found in plant tissues in the form of phenolic acids, free and glycosylated flavonoids, and anthocyanins that are a subgroup of flavonoids [3].

The seeds of the plant Trigonella contain about 1–3% polyphenols [14], which are found in many forms, most notably glycosylated flavonoids [15, 16]. The plant has been known since ancient times for its beneficial properties both in human health and in its applications in increasing food preservation [3]. Regarding the phenolic components found in the seeds of the Trigonella plant, vanillic acid, 3-coumaric acid, genetic acid, and caffeic acid show larger amounts. Their amounts are, respectively, 0.585, 0.478, 0.358, and 0.210 mg/g seed [17–19]. However, other types of phenolics have been identified in the seed of the plant that are not mentioned as they are not, quantitatively, a significant percentage. In addition, the total phenolic content of the seeds ranges from 10 to 78 mg GAE/g, and the variation depends on the extraction and collection technique of the total phenolic components.

However, the seed of the Trigonella plant also contains a significant percentage of flavonoids, which, however, are mostly in glycosylated forms. In particular, the main aglycones contained in the seeds are apigenin with its glycosides accounting for about 40% of the total amount of flavonoids, camphorol with glycosides constituting about 15% of the total amount of flavonoids, and luteolin with glycosides constituting about 15% of the total amount of flavonoids. However, some of these glycosides have been identified for the first time in this plant, such as vicenin and its derivatives, which are xylose glycosides of apigenin, orientine, which is a glycoside of luteolin, and vitexin, which is a glycoside of apigenin [15–20].

One of the ingredients added to meat emulsion products is olive oil, which is usually added to replace pork fat, in order to develop a healthy fatty acid profile, in the presence of monounsaturated fatty acids derived from olive oil. In meat mass, the application methods that have been proposed at research and industrial levels include the direct addition of olive oil in liquid form during processing in the cutter [21] and the pre-emulsification of olive oil [6]. The olive oil added during the production of cooked sausages must comply with the specifications of European Regulation 2568/912013, which sets the limits of the composition of fatty acids to qualify an oil as olive oil and, in particular, must contain 70–80% monounsaturated fatty acids (oleic acid), 6–16% omega-6 fatty acids, 0.3–1.3% omega-3 fatty acids, and 8–10% saturated fatty acids. In addition, in terms of organoleptic characteristics, olive oil used in the production of cooked sausages should have a fruity taste and smell as well as not show unacceptable organoleptic characteristics of taste and smell that refer to an oxidized product. The color of olive oil, which is influenced by factors such as storage conditions, place of production, and export system [22], is a result of chlorophylls where they are green in color as well as carotenoids (β-carotene and lutein) where they show yellow coloration. In meat products and especially in cooked sausages, olive oils rich in carotene and not in chlorophyll are preferable, as the development of green colors in the final products is considered unacceptable [23, 24].

The scope of this study was to evaluate the color preservation effect of T. foenum-graecum seed powder in meat emulsion systems and the potential replacement of nitrites with the specific seed powder.

2. Materials and methods

2.1 Materials

Fresh minced pork ham (*M. biceps femoris, M. semitendinosus, M. semimembranosus*) was purchased from a local market at 48-h postmortem. T. foenum-graecum seeds

Ingredient	Quantity (g)	
	Formulation 1—Control	Formulation 2
Beef minced meat	274.3	274.3
Water-Ice	125.7	125.7
Olive oil	75	75
Sodium chloride	10	10
Dtfg	—	15
Starch	15	—
Sodium nitrite ($NaNO_2$)	0.075	
Sum	500	500

Table 1.
Meat emulsion formulation.

were purchased from a local market, and then, it was grounded using a laboratory grinder (Analytische Mühle, IKA). Sodium chloride (Kallas klassiko), corn starch (Bioygeia), and virgin olive oil (Altis Klassiko, ELAIS) were purchased from local markets. Sodium nitrite was obtained from CG Chemicalien (CG Chemicalien, Belgium).

2.2 Meat emulsion preparation

The meat emulsions were prepared on the basis of complete replacement of sodium nitrite and starch with Trigonella seed powder, where the fat was removed by the Soxhlet method. Thus, two samples were manufactured, namely, the first control sample containing 3% starch and 150 ppm sodium nitrite, which is the upper limit according to European Regulation 1129/2011 on food additives. The second sample contained 3% deffated T. foenum-graecum seed powder, where the fat was removed by the Soxhlet method without the presence of sodium nitrite. The recipes of meat emulsions are presented in **Table 1**.

2.3 Estimation of levels of oxymyoglobin Mb (Fe^{2+}) and metamyoglobin MetMb (Fe^{3+}) of fresh meat emulsions

The estimation of the levels of oxymyoglobin Mb (Fe^{2+}) and metamyoglobin MetMb (Fe^{3+}) in fresh meat emulsions was based on the method of Ning et al. [21] and Carlez et al. [22] with some modifications. Specifically, 2 g of the sample was homogenized with 20 ml of Na/K phosphate buffer at a concentration of 0.04 mol/L in a 50-ml Falcon flask in an Ultraturrax homogenizer for 20 s at 10,800 rpm. The samples were then centrifuged at 7000 rpm for 5 min after a total of 1 h in an ice basin, with the addition of aluminum foil externally to prevent oxidation. Immediately after filtration with whatman #1 filter type, the samples were supplemented with 25 ml of the same buffer. The absorbance of the samples at the following wavelengths was then measured: 525, 545, 565, and 572 nm, in a spectrophotometer (UV-1800, Shimadzu Co., Kyoto, Japan). The percentages of oxymyoglobin and metamyoglobin are calculated from the following equations [25]:

$$MbO_2\% = (0.882 * R_1 - 1.267 * R_2 + 0.809 * R_3 - 0.361) * 100 \qquad (1)$$

$$\text{MetMb\%} = (-2.541 * R_1 + 0.777 * R_2 + 0.800 * R_3 + 1.098) * 100 \tag{2}$$

where factors R_1, R_2, and R_3 are the absorption ratios and are calculated as follows:

$$R_1 = \frac{A572}{A525}, R_2 = \frac{A565}{A525}, R_3 = \frac{A545}{A525}. \tag{3}$$

2.4 Deffating

The defatting of T. foenum-graecum seeds was carried out based on the method of lipid extraction [26]. In this method, for the defatting of raw seeds, a Soxhlet apparatus was used utilizing diethylether as solvent. More specifically, a certain amount of T. foenum-graecum seed powder was weighed and placed in a cellulose extraction cartridge. The cartridge was plugged with cotton wool and then placed in the Soxhlet chamber, which was fitted to a pretared distillation flask containing dielthylether and two to three glass regulators. After extraction for 2 h in 50–60°C, the cartridge allowed to cool, and then, the cartridges were placed in an oven at 105°C for 12 h, following cooling in a desiccator and weighing. Placing in the oven for specific time, cooling in a desiccator and weighing was repeated until the difference between two consecutive weights was smaller than 2 mg.

2.5 Determination of color parameters

Instrumental color was conducted by taking a direct reading of raw meat emulsions with a colorimeter (Konica Minolta CR-400, USA). Standard observer was 2° (Closely matches CIE 1931 Standard Observer [(x2λ, yλ, zλ)]); 8-mm diameter circular aperture and d/0 (D65,Diffuse illumination/0 < ° viewing angle) illuminate were used. The CIE-L^*, a^*, and b^* parameters were evaluated according to the methodology proposed by the American Meat Science Association (AMSA 2012). The calibration of colorimeter performed using a white plate, $L^* = 97.58$, $a^* = 0.03$, and $b^* = 1.08$. For the color analysis of raw meat emulsions, 10 g of each sample was placed on a 9-cm diameter petri dish at 1-cm thickness. The average of nine readings was reported. All tests were performed at room temperature on the second day after manufacture and then 5 and 7 days after manufacturing.

2.6 Statistical analysis

All statistical analyses were performed with SPSS (Version 23, IBM, USA). The homogeneity of variances was tested with the Levene test in the case of non-normal data with SPSS. The normality of the residuals was tested using UNIVARIATE of SPSS, and consideration was given to the Shapiro–Wilk test for normality. Once it was determined that the assumptions of the analysis of variance (ANOVA) were met for these data, the GLM procedure of SAS with a fixed effect of treatment and a random effect of replication was used for the statistical determination of all variables. Tukey's test *($P < 0.05$)* was used to determine the differences between the treatment means. Least squares means were separated using a single-degree-of-freedom estimate statement to determine the difference between meaningful comparisons, which included Starch + $NaNO_2$ versus Dtfg (3% inclusion). Differences were considered statistically different at $P < 0.05$. This experiment was completed in its entirety three times.

3. Results and discussion

3.1 Estimation of levels of oxymyoglobin Mb (Fe^{2+}) and metamyoglobin MetMb (Fe^{3+}) of raw meat emulsions and determination of color parameters

In **Figures 1–4**, it is observed that in the Starch + $NaNO_2$ sample, the oxymyoglobin levels are lower than in the Dtfg sample throughout the shelf life of the meat emulsions. In addition, in the Starch + $NaNO_2$ sample, lower metamyoglobin values were observed after the fifth day of maintenance. These results could be correlated with the analysis of color parameters, as shown in **Figure 5**, where of particular

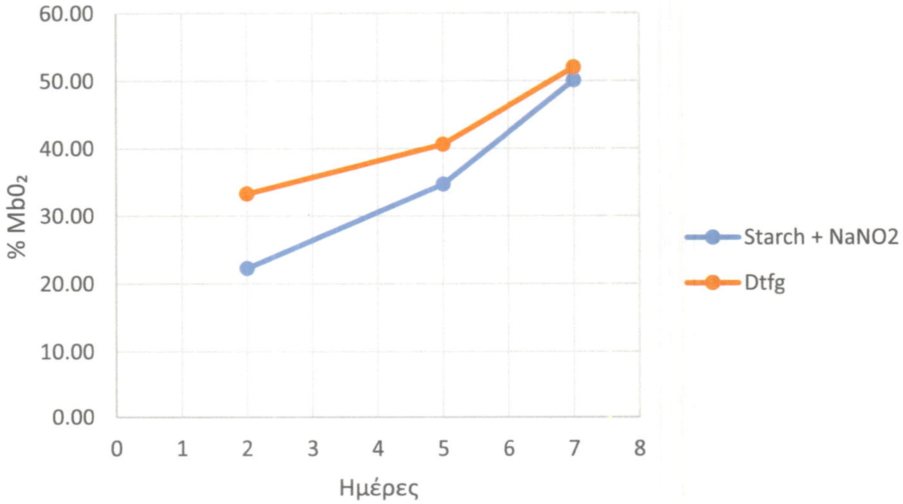

Figure 1.
Effect of deffated Trigonella foenum-graecum seed powder in oxymyoglobin content of meat emulsions with olive oil.

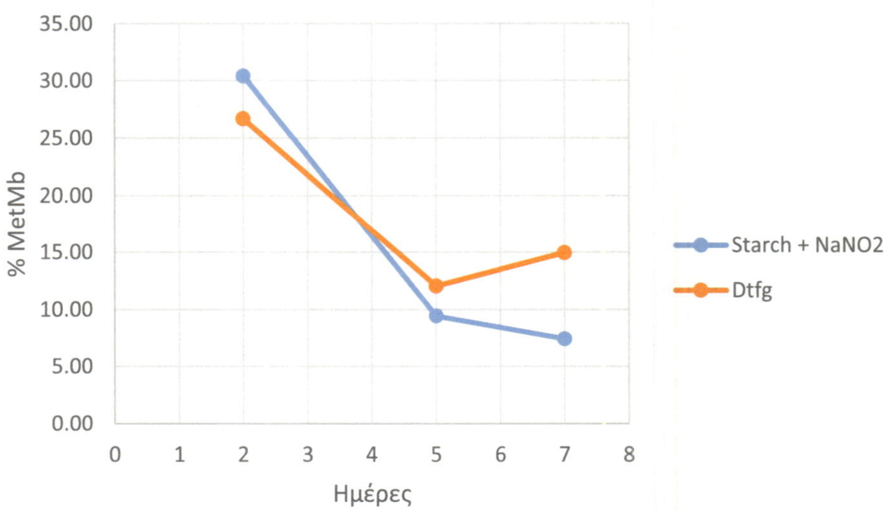

Figure 2.
Effect of deffated Trigonella foenum-graecum seed powder in metmyoglobin content of meat emulsions with olive oil.

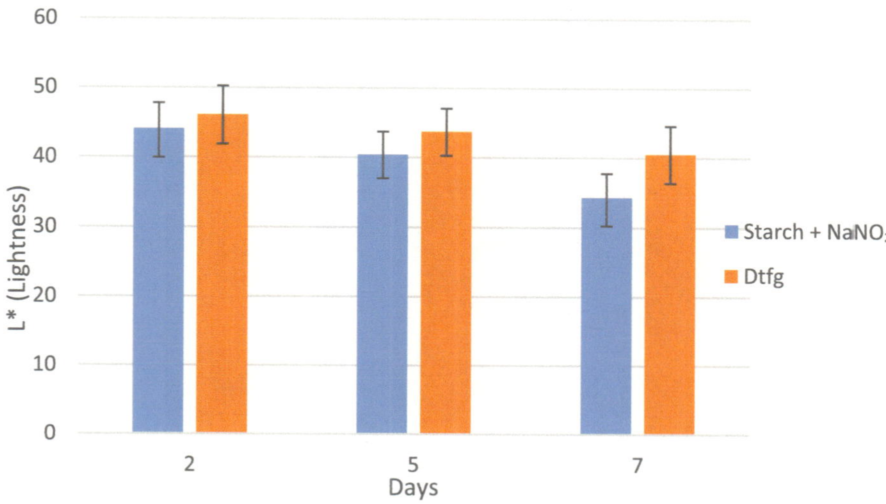

Figure 3.
Effect of deffated Trigonella foenum-graecum seed powder in L (lightness) factor of meat emulsions with olive oil.*

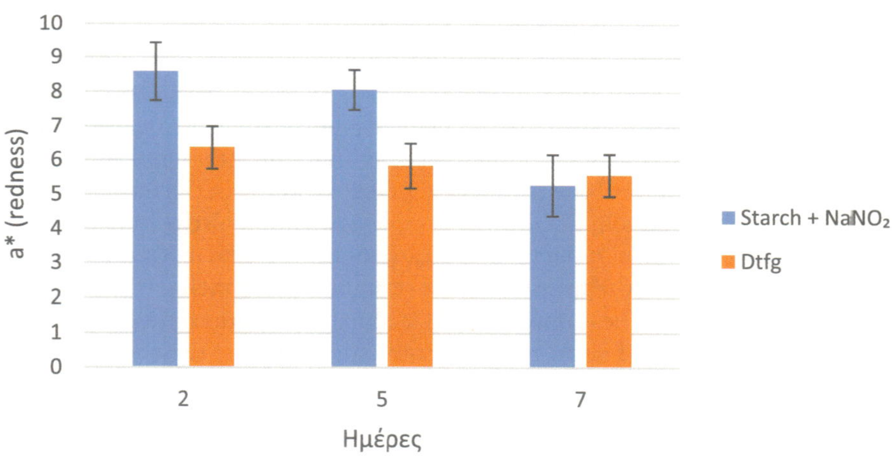

Figure 4.
Effect of deffated Trigonella foenum-graecum seed powder in a (redness) factor of meat emulsions with olive oil.*

interest is the sharp reduction of the factor L^* (brightness) and a^* (red color) in the Starch + $NaNO_2$ sample, in contrast to the Dtfg sample.

These findings are particularly important; as in the present literature, there is difficulty in implementing strategies for the total replacement of nitrites with different plant materials. However, in some studies, an increase in color stability has been observed with the use of plant materials in meat products low in nitrite [12]. The stabilization of the color of meat products with the use of Trigonella plant powder is probably due to its high content of antioxidants, such as phenolic compounds and some steroids [27], as similar effects are observed during the incorporation of essential oils, with high content of antioxidants, in meat products with low nitrate content [26, 28]. These indications may be an important tool for the use of the Trigonella plant in the development of strategies for the reduction and replacement of nitrites in meat products, especially given that the samples of the present work were kept for 7 days without packaging in an environment with high redox potential without undergoing heat treatment.

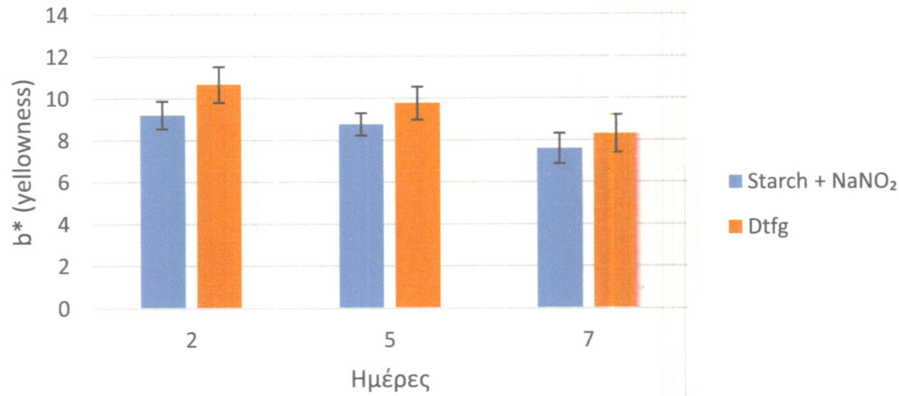

Figure 5.
Effect of deffated Trigonella foenum-graecum seed powder in b (yellowness) factor of meat emulsions with olive oil.*

4. Conclusion

The replacement of nitrates and starch from Trigonella seed powder in beef emulsions with olive oil, the stability of myoglobin to oxidation, and certain color factors were evaluated. Thus, the percentage of oxymyoglobin in the meat emulsion containing Trigonella was higher than in the meat emulsion containing nitrites and starch. In addition, the percentage of metamyoglobin in the meat emulsion containing Trigonella was lower until the fifth day of preservation than the meat emulsion containing nitrites and starch. Finally, the sample containing nitrites and starch underwent a sharp decrease in the values of L^* (brightness) and a^* (red color) during maintenance compared to the meat emulsion containing Trigonella, which constitutes its highest color stability. These observations lead to the hypothesis that T. foenum-graecum seed powder could be an efficient candidate for nitrite replacement from a color stability point of view in meat emulsion systems with high olive oil content.

Author details

Theofilos Frangopoulos
International Hellenic University, Thessaloniki, Greece

*Address all correspondence to: thfrangopoulos@gmail.com

References

[1] Flammang A, Cifone M, Erexson G, Stankowski L. Genotoxicity testing of a fenugreek extract. Food Chemistry and Toxicology. 2004;**11**:1769-1775

[2] Hutchinson J. The Genera of Flowering Plants. Vol. 1. Oxford: Clarendon Press; 1964

[3] Ahmad A, Alghamdi SS, Mahmood K, Afzal M. Fenugreek a multipurpose crop: Potentialities and improvements. Saudi Journal of Biological Sciences. 2016;**23**:300-310. DOI: 10.1016/j.sjbs.2015.09.015

[4] Miller JI. The Spice Trade of the Roman Empire 29 B.C. to A.D. 641. Oxford: Clarendon Press; 1969

[5] ICAP, 2015

[6] Kouzounis D, Lazaridou A, Katsanidis E. Partial replacement of animal fat by oleogels structured with monoglycerides and phytosterols in frankfurter sausages. Meat Science. 2017;**130**:38-46. DOI: 10.1016/j.meatsci.2017.04.004

[7] Jiménez-Colmenero F, Herrero A, Pintado T, Solas MT, Ruiz-Capillas C. Influence of emulsified olive oil stabilizing system used for pork backfat replacement in frankfurters. Food Research International. 2010;**43**:2068-2076. DOI: 10.1016/j.foodres.2010.06.010

[8] Youssef MK, Wang Q, Cui SW, Barbut S. Purification and partial physicochemical characteristics of protein free fenugreek gums. Food Hydrocolloids. 2009;**23**:2049-2053. DOI: 10.1016/j.foodhyd.2009.03.017

[9] Herrmann SS, Granby K, Duedahl-Olesen L. Formation and mitigation of N-nitrosamines in nitrite preserved cooked sausages. Food Chemistry. 2015;**174**:516-526. DOI: 10.1016/j.foodchem.2014.11.101

[10] Sucu C, Turp GY. The investigation of the use of beetroot powder in Turkish fermented beef sausage (sucuk) as nitrite alternative. Meat Science. 2018;**140**:158-166. DOI: 10.1016/j.meatsci.2018.03.012

[11] Jafari M, Emam-Djomeh Z. Reducing nitrite content in hot dogs by hurdle technology. Food Control. 2007;**18**:1488-1493. DOI: 10.1016/j.foodcont.2006.11.007

[12] Ruiz-Capillas C, Tahmouzi S, Triki M, Rodriguez-Salas L, Jimenez-Colmenero F, Herrero AM. Nitrite-free Asian hot dog sausages reformulated with nitrite replacers. Journal of Food Science and Technology. 2015;**52**:4333-4341. DOI: 10.1007/s13197-014-1460-1

[13] Wani SA, Kumar P. Fenugreek: A review on its nutraceutical properties and utilization in various food products. Journal of the Saudi Society of Agricultural Sciences. 2018;**17**:97-106. DOI: 10.1016/j.jssas.2016.01.007

[14] Benayad Z, Gomez-Cordoves C, Es-Safi NE. Characterization of flavonoid glycosides from fenugreek (Trigonella foenum-graecum) crude seeds by HPLC-DAD-ESI/MS analysis. International Journal of Molecular Sciences. 2014;**15**:20668-20685. DOI: 10.3390/ijms151120668

[15] Benayad Z, Gómez-Cordovés C, Es-Safi NE. Identification and quantification of flavonoid glycosides from fenugreek (Trigonella foenum-graecum) germinated seeds by

LC–DAD–ESI/MS analysis. Journal of Food Composition and Analysis. 2014;**35**: 21-29. DOI: 10.1016/j.jfca.2014.04.002

[16] Rababah TM, Ereifej KI, Esoh RB, Al-u'datt MH, Alrababah MA, Yang W. Antioxidant activities, total phenolics and HPLC analyses of the phenolic compounds of extracts from common Mediterranean plants. Natural Product Research. 2011;**25**:596-605. DOI: 10.1080/14786419.2010.488232

[17] Pajak P, Socha R, Broniek J, Krolikowska K, Fortuna T. Antioxidant properties, phenolic and mineral composition of germinated chia, golden flax, evening primrose, phacelia and fenugreek. Food Chemistry. 2019;**275**:69-76. DOI: 10.1016/j.foodchem.2018.09.081

[18] Belguith-Hadriche O, Bouaziz M, Jamoussi K, Simmonds MS, El Feki A, Makni-Ayedi F. Comparative study on hypocholesterolemic and antioxidant activities of various extracts of fenugreek seeds. Food Chemistry. 2013;**138**:1448-1453. DOI: 10.1016/j.foodchem.2012.11.003

[19] Kenny O, Smyth TJ, Hewage CM, Brunton NP. Antioxidant properties and quantitative UPLC-MS analysis of phenolic compounds from extracts of fenugreek (Trigonella foenum-graecum) seeds and bitter melon (*Momordica charantia*) fruit. Food Chemistry. 2013;**141**:4295-4302. DOI: 10.1016/j.foodchem.2013.07.016

[20] Pappa IC, Bloukas JG, Arvanitoyannis IS. Optimization of salt, olive oil and pectin level for low-fat frankfurters produced by replacing pork backfat with olive oil. Meat Science. 2000;**56**:81-88. DOI: 10.1016/S0309-1740(00)00024-3

[21] Ning C, Li L, Fang H, Ma F, Tang Y. & Zhou C.l-lysine/l-arginine/l-cysteine synergistically improves the color of cured sausage with $NaNO_2$ by hindering myoglobin oxidation and promoting nitrosylmyoglobin formation. Food Chemistry. 2019;**284**:219-226. DOI: 10.1016/j.foodchem.2019.01.116

[22] Carlez A, Veciana-Nogues T, Cheftel J-C. Changes in colour and myoglobin of minced beef meat due to high pressure processing. LWT—Food Science and Technology. 1995;**28**:528-538. DOI: 10.1006/fstl.1995.0088

[23] Aburezq HA, Mansour M, Safer A-M, Afzal M. Cyto-protective and immunomodulating effect of Curcuma longa in Wistar rats subjected to carbon tetrachloride-induced oxidative stress. Inflammopharmacology. 2008;**16**:87-95. DOI: 10.1007/s10787-007-1621-1

[24] Kim HW, Miller DK, Lee YJ, Kim YH. Effects of soy hull pectin and insoluble fiber on physicochemical and oxidative characteristics of fresh and frozen/thawed beef patties. Meat Science. 2016;**117**:63-67. DOI: 10.1016/j.meatsci.2016.02.035

[25] Krzywicki K. Assessment of relative content of myoglobin, oxymyoglobin and metmyoglobin at the surface of beef. Meat Science. 1979;**3**:1-10

[26] Vafania B, Fathi M, Soleimanian-Zad S. Nanoencapsulation of thyme essential oil in chitosan-gelatin nanofibers by nozzle-less electrospinning and their application to reduce nitrite in sausages. Food and Bioproducts Processing. 2019;**116**:240-248. DOI: 10.1016/j.fbp.2019.06.001

[27] Frangopoulos T. Incorporation of Trigonella foenum-graecum seed powder in meat emulsion systems with olive oil: Effects on physicochemical, texture, and color characteristics. Journal of

Food Science and Technology. 2021. DOI: 10.1007/s13197-021-05220-3

[28] Shin DM, Hwang KE, Lee CW, Kim TK, Park YS, Han SG. Effect of Swiss chard (*Beta vulgaris* var. *cicla*) as nitrite replacement on color stability and shelf-life of cooked pork patties during refrigerated storage. Korean Journal for Food Science of Animal Resources. 2017;**37**:418-428. DOI: 10.5851/kosfa.2017.37.3.418

Chapter 5

Sensory Evaluation of EVOO: Do Different Test Locations Have a Relevant Impact on Data Quality?

Annette Bongartz, Martin Popp and Richard Retsch

Abstract

In natural sciences, in general, the most important challenge is to ensure the reliability and validity of collected data and results. The identification of relevant influencing factors and the definition of adequate methodological approaches helps to minimize "noise." Within the sensory evaluation of extra virgin olive oil, many potentially influencing factors are known. Tasting procedures, therefore, are standardized. However, not all criteria have the same impact on data quality. The study at hand focuses on the possible influence of different test locations, comparing the situation in sensory labs (*in situ*) with so-called home testing stations (remote) for two well-experienced olive oil panels. Panel performance of both panels meets all regulatory requirements. Looking at the results from the overall statistical data analysis, slight differences between results coming from the two panels can be seen (nevertheless, not exceeding the requirements), but almost no differences are found between results coming out of different test situations. Knowing that the influence of testing through different panels is small but nevertheless bigger than a potential impact of testing in different test locations (sensory lab versus home testing stations), shows us the great potential for future use of remote test designs likewise to lab designs to obtain valid data.

Keywords: sensory evaluation, test location, remote testing, extra virgin olive oil (EVOO), data validity

1. Introduction

In natural sciences, the standardization of operation procedures, aiming at methodologies and the reliability as well as the validity of resulting data, is most important.

Thinking of test settings in sensory science, especially to control and monitor panelist and panel performance during the panel work in defined test settings is necessary. In the context of sensory evaluation of olive oil specifically, several official requirements do exist—on the one hand the EEC regulation 2568/91 [1], as amended, as well as several underlying documents and guidelines from the International Olive Council (IOC) [2–4] and moreover the general EN ISO/IEC 17025 regulations for any kind of testing laboratories [5].

Data quality must be the overriding objective in natural sciences and therefore is indispensable. Assuring a high data quality during data collection and assessment requires a clear focus on "data reliability" (high precision → same/similar results) and "data validity" (high accuracy → correct results). Well known is that one can gain high precision in measuring something, but at the same time can miss the target—meaning that results are precise, but not correct. So, overall high data quality can only be achieved, if data are on the one hand precise (reliable) and at the same time as well accurate (valid).

- Reliability: consistency and precision of measurement (in sensory analysis—repeatability of results—over time, of single panelists, of whole panels, indifferent test locations, ...)

- Validity: accuracy and trueness of measurement (in sensory analysis—homogeneity of results—between panels, in different test locations, ...)

Factors that might have a negative impact on data quality in sensory science are manifold. They can be related to the execution of the general procedure (test methodology), to the handling of test samples (blinding, distribution, temperature), to training and monitoring aspects of panelists and panels, to statistical analysis and data management as well as to the test infrastructure.

All above-named regulations and guidelines have in common to standardize and control procedures and finally to minimize "noise" in resulting data. This is helpful and valuable, but nevertheless, not all possible and focused so-called "influencing factors" cause a similar or even a relevant impact on data quality—some of them, presupposing a specific framework of instructions and settings, even have none.

2. Background and objective

The study at hand focuses on the aspect of different test locations for objective sensory analysis and their possible impact on the quality of resulting data.

Normally tests in sensory analysis take place as central location tests (CLT) in standardized testing rooms, such as sensory laboratories. Most of the regulations and guidelines propose these "*in situ*" (on-site) approaches. Panelists in these cases come to the lab, use the provided infrastructure (test cabins, hard and software for data collection, maybe additional technical equipment, such as heaters) for training and testing and at the same time, all other surrounding conditions (e.g., light, climate (humidity, temperature), else) are defined and controlled automatically. This is normally the rule and easily to be organized when using so-called "internal panels," consisting of panelists that work nearby, in the same institution, company, etc. Panelists can be scheduled quite easily and are able to take part in either regularly planned or as well spontaneously organized synchronic test situations.

But how about panelists in so-called "external panels," who must travel to be able to participate in regular on-site and synchronic trainings and test situations? Such dates can be organized and scheduled only medium or long-term presupposed. Rather flexible and spontaneous testing under such conditions is almost not possible. This consideration shows us—independently from pandemic situations—the need for additional appropriate test settings, that on the one hand can secure high standards of data quality/validity in sensory analysis and are on the other hand flexible as well as time and cost-efficient. Doing sensory trainings and tests with panelists "remote" (off-site), meaning that panelists work at home respectively at defined and standardized "home-testing-stations," cause less costs for traveling (time per panelist, transportation). Data can be collected and exchanged online (web-based) either synchronously or asynchronously and as well panel meetings can take place remotely. Moreover, the latest data even show a better availability of panelists, due to less necessary time effort for traveling and thereby as well a higher motivation for contribution in panel work, compared to more elaborate (*in situ*) settings [6].

The aim of the study at hand is to prove the overall performance of participating sensory panels and the quality of the collected data. The focus lies on the evaluation of the suitability of "*in situ*" test situations (sensory laboratory) versus "remote" test situations (home test stations) for the sensory evaluation of olive oil. Additionally, the criteria of homogeneity between the considered panels as well as homogeneity, consistency, and repeatability of each single panel are particularly interesting in this context.

3. Materials and methods

3.1 Panels and panelists

The study at hand compares results from two sensory olive oil panels, namely the German Olive Oil Panel/DOP and the Swiss Olive Oil Panel/SOP.

Both panels are objective expert panels whose members have many years of experience in the sensory evaluation of olive oil. Specific infrastructure makes it possible to either carry out sensory tests synchronously and "*in situ*" in sensory laboratory (or comparable) situations, equipped with test booth, heaters, etc. On the other hand, both panels have the possibility to test asynchronously at so-called home testing stations. All panelists from both panels are regularly trained and undergo permanent monitoring of their panelist and panel performance. Additionally, both panels take part in regular inter-laboratory comparisons (proficiency tests). This is—in addition to other requirements—an important basis for their consistent, reliable, and successful work. Engaged in research projects as well as services, both panels contribute to the continuous development of quality on the olive oil market.

3.1.1 Swiss olive oil panel (SOP)

The SOP consists overall of 38 panelists and was founded in 2002. Since 2006 the SOP is accredited in accordance with EN ISO/IEC 17025 [5] and has been recognized by the International Olive Council (IOC) between 2009 and 2021. In the study at hand, the same nine panelists contributed "*in situ*" and synchronously to a test in a sensory laboratory situation as well as "remote" and asynchronously at their home testing stations.

3.1.2 German olive oil panel (DOP)

The DOP consists overall of 25 panelists and was founded in 1999. Since 2012 the DOP is accredited in accordance with EN ISO/IEC 17025 [5] and has been recognized by the International Olive Council (IOC) between 2012 and 2021. In the study at hand, the same 11 panelists contributed "*in situ*" and synchronously to a test in a sensory laboratory situation and "remote" and asynchronously at their home testing stations.

3.2 Test situations

There are at least two options to conduct sensory tests—on the one hand "*in situ*" the more common version of central location tests (CLT) in standardized testing rooms, for example, sensory laboratories, where panelists work synchronously and on the other hand "remote" the—at least in sensory analysis—less common home tests (HT) at so-called home testing stations, meaning in panelists homes, where panelists work in personal, but nevertheless standardized workspaces. Thinking about the advantages and disadvantages of the named test situations, we find pros and cons for both sides.

*3.2.1 Sensory laboratories (*in situ*)*

The sensory laboratory of SOP is located on the ZHAW campus in Wädenswil. It consists of 12 separate test booths. Each booth is equipped with a computer (incl. data acquisition software FIZZ by Biosystemes) and a heating device by Ettore Pasquali (mod. 145). Each heating device is recorded in a device list of the QMS and is checked regularly. Detailed information concerning the procedure of testing in the sensory laboratory (and concerning additional equipment, such as test glasses, spittoons, and palate-cleansing agents) is described in the QMS (internal document: Standard Operation Procedure: LMT-SEN-A5-302_translated EN → Sensory Evaluation at ZHAW (Sensory Laboratory)).

The sensory laboratory of DOP is in Nuremberg, Maxfeldstrasse 50. It consists of 12 separate mobile test booths. Each booth is equipped with a computer (incl. data acquisition software SENSORY by IMEDIA) and a heating device by Ettore Pasquali (mod. 145). Each heating device is recorded in a device list of the QMS and is checked regularly. Detailed information can be found in the QMS (internal document: Standard Operation Procedure SOP 07-02-02).

On the "Pro" side (**Table 1**) there is of course a high level of standardization. Samples are prepared in an absolute neutral way by the panel leader (or a technician), the testing takes place synchronously (at the same time) and electronic equipment is used for data collection and analysis.

On the "Contra" side (**Table 1**) we see low flexibility in terms of scheduling tests, especially because panelists must be available at the same time. If panelists are not collaborators and work on-site, they must travel. This is time- (and cost-) consuming.

Pros	Cons
ISO 17025 accreditation leads to a high level of standardization (controlled test situation, panelist/panel performance, etc.)	Synchronous testing causes low flexibility in terms of timing/scheduling tests → all testers have to be available at the same time
Sample preparation (pouring 15 ml) in coded glasses is done by the panel leader	Necessity for testers to get on-site requires traveling-time and is rather time-consuming for them
Synchronous testing (8–12 panelists)	Other potential influencing factors (malfunction interruption, noise, etc.) ...
Use of electronic equipment to collect, compile and analyze data	

Table 1.
Pros/cons of a test situation in a sensory laboratory.

Moreover, even in a sensory lab, there is potential for additional interruptions (noise, malfunction, or else) with negative effects.

3.2.2 Home testing stations (remote)

Within the SOP, each panelist has set up a personal testing station in a room in his or her home. Each testing station is equipped with a heating device (Rosenstein & Söhne), a thermometer (Testo Mini penetration thermometer), and blue glasses including cover glasses. Each heating device and the thermometer is recorded in a device list of the QMS and is checked regularly. Detailed information concerning the procedure of testing in a home testing station is described in the QMS (internal document: Standard Operation Procedure: LMT-SEN-A5-303_translated EN → Sensory Evaluation at Home Testing Stations).

Within the DOP each panelist has set up a personal testing station in a room in his or her home, which meets the test conditions of the IOC in terms of light, temperature, noise, and odors (COI/T.20/DOC. No 6/Rev. 1). The mobile booth is made up of folding elements in such a way that the panelist is isolated from negative ambient conditions. Each test station is equipped with a heating device (Ettore Pasquali, mod. 145), a thermometer (Testo Mini penetration thermometer), and blue glasses, including a cover glass. Each heating device and the thermometer is recorded in a device list of the QMS and is checked regularly. Detailed information can be found in the QMS (internal document: Standard Operation Procedure SOP 07-02-02).

On the "Pro" side (**Table 2**) there is—like for a lab situation—as well a high level of standardization because as well home testing stations provide ISO accreditation. Asynchronous testing increases flexibility for scheduling tests, meaning that testers do not have to be available at the same time, only in a defined period. Less traveling time and costs are positive and of course—like in the lab situation—as well at home testing stations, electronic equipment for data collection and analysis is used.

On the "Contra" side (**Table 2**), we see that a sample dispatch is needed, which must be well organized regarding packaging and preparation of samples. For temperature protection during transport, Styrofoam boxes are used, and the oil is coded

Pros	Cons
ISO 17025 accreditation leads to a high level of standardization → controlled test situation, panelist/panel performance, etc.	Sample dispatch is needed → 30 ml (per olive oil) in dark glass bottles and use of styroporous boxes which secure sample temperature
Asynchronous testing enables high flexibility in terms of timing/scheduling tests → testers do not have to be available at the same, they have to respect deadlines, but otherwise can plan rather independently	Sample preparation (pouring 15 ml) in coded glass is done by each panelist → half of the whole 30 ml quantum
	Asynchronous testing (8–12 panelists) → no relevance because of training status
Less requirement of traveling-time because of remote testing (at home)	Other potential influencing factors (malfunction interruption, noise, etc.) ...
Use of electronic equipment to collect, compile and analyze data	

Table 2.
Pros/cons of a test situation at a home testing station.

and bottled to dark 30 ml bottles. Sample preparation, directly before testing, is done by the panelist. He or she must pour out exactly half of the bottle (15 from 30 ml). Asynchronous testing is of course different compared to asynchronous lab testing situation, but since panelists are well trained on using the methodology and it is always (as well in the lab) a single panelist evaluation before compiling data, this does not lead to any problems. Finally, and again like in the lab situation, there is of course potential for additional interruptions (noise, malfunction, or else).

3.3 Test methodology (EU 2568/91, as amended)

The applied sensory methodology is based on the official panel test according to the regulation EEC regulation 2568/91 [1] and related IOC documents.

3.4 Validation concept

To be able to record high data quality in the context of sensory evaluation of olive oil, the reliability and validity of raw data must be ensured. Therefore, a study concept, based on the recommendations of the IOC (COI/T.28/Doc. No.1/Rev. 5 2019), was considered. Among other criteria, analyzing the panelist and panel performance, the concept focuses especially on the aspect of the test situation (sensory laboratory versus home testing stations) (**Table 3**).

3.5 Test samples

Both participating panels did evaluate the same selection of 10 olive oils (same lot number) "*in situ*" (in the sensory laboratory) and "remote" (at home testing stations), see **Table 4**. The order of testing was the same in all test situations but in each test situation individual three-digit codes were used to avoid influencing the testers.

	Test situation	Panel	Panel performance
Focus	Sensory laboratory vs. Home testing stations	DOP vs. SOP	• Homogeneity • Consistency • Repeatability
Attributes	• Fruitiness • Bitterness • Pungency • Defect	• Fruitiness • Bitterness • Pungency • Defect	• Fruitiness • Bitterness • Pungency • *Defect
Methodological approach	• Graphical visualization • Mixed-model ANOVA • Analyzed for each panel separately	• Graphical visualization • Mixed-model ANOVA	• Trueness (<2.0) 1. *z-Score → homogeneity* 2. *Deviation (DNp)* from other panels per session → homogeneity • Precision (<2.0) 1. *Normalized error (En) — panel mean → repeatability* 2. *Precision number (PNp) — panel mean → consistency* • Analyzed for each panel separately

*not analyzed in present study

Table 3.
Validation criteria.

No.	Sample code*	Product information
P1	104	100% Italian olive oil; different varieties, extra virgin
P2	507	European Blend; different varieties, extra virgin
P3	620	100% Italian olive oil, 100% Nocellara, extra virgin
P4	733	100% Italian olive oil; different varieties, extra virgin
P5	249	100% French olive oil; different varieties, extra virgin
P6	362	Olive Oil from IOC Org 2—2020, extra virgin
P7	878	Olive Oil from IOC Org 1—2020, defective
P8	168	Olive Oil from IOC Org 1—2020, extra virgin
P9	055	Olive Oil from IOC Org 2—2020, defective
P10	652	Olive Oil from IOC Org 2—2020, defective

For each test situation, individual three-digit codes were used to avoid influencing the testers.

Table 4.
Test samples.

3.6 Data collection

Both participating panels did evaluate the test samples in the same period, but independently. The evaluation criteria on the used profile sheets from both panels (electronically/paper) were identical, corresponding with the EEC regulation 2568/91 [1].

3.6.1 Swiss olive oil panel/SOP

To collect data in the sensory laboratory of ZHAW, panelists from SOP are provided with PC's in the test booth, equipped with the sensory software "Fizz" (Biosystemes Fizz for Windows 2.46 A), which allows direct electronic recording of individual panelist data on an electronic profile sheet.

For the collection of data at home testing stations, panelists from SOP use a profile sheet (paper) and transfer individual panelist data, online via the internet, to the panel leader, using the software "LimeSurvey."

3.6.2 German olive oil panel/DOP

To collect data in the sensory laboratory situation, panelists from DOP are provided with PC's in the test booth, equipped with the software "SENSORY" (IMEDIA), which allows direct electronic recording of individual panelist data on an electronic profile sheet.

For the collection of data at home testing stations, panelists from DOP use a profile sheet (paper) and transfer individual panelist data, online via the internet, to the panel leader, using the software "SENSORY" (IMEDIA).

3.7 Data evaluation (assessment)

All results of the different tasting sessions with all panelists and all panels were combined to a common data set, using the software program "Excel" (Microsoft

Office Excel 365). The following data evaluation was done with help of the Add-in Software "XLStat" (version 2020).

4. Results and discussion

In the study at hand altogether 10 olive oils (seven of them extra virgin and three defective ones) were independently tested by the Swiss Olive Oil Panel/SOP (nine panelists) and the German Olive Oil Panel/DOP (11 panelists) in different test situations—namely in sensory laboratories (*in situ*) and at so-called home testing stations (remote). Test results were analyzed regarding the validity of the data and thereby focusing on the following three aspects:

1. Agreement between test situation (*in situ* versus remote)

2. Agreement (homogeneity) between panels (SOP versus DOP)

3. Individual performance of both panels (SOP, DOP)

All data/panel results were valid according to IOC specifications (e.g., Cvr < 20%).

4.1 Agreement between different test situations (*in situ* versus remote)

First, and for both panels separately, the agreement between data collected in different test situations—sensory laboratory (*in situ*) and home testing stations (remote testing)—was checked. For this purpose, raw data were analyzed using mixed-model ANOVA. Results show that there are no statistically significant differences between

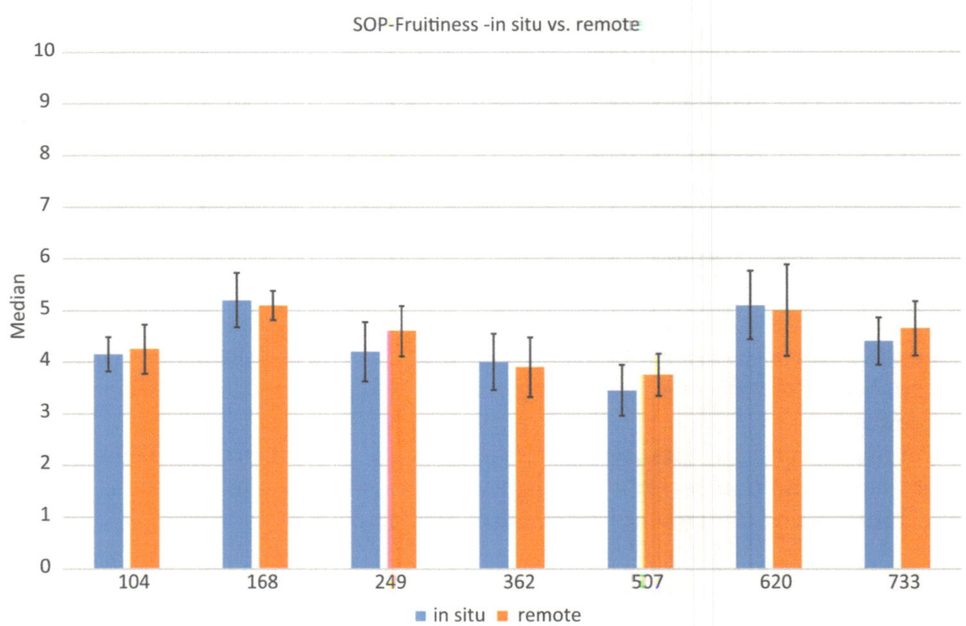

Figure 1.
Median of fruitiness (SOP)—in situ versus remote (n = 9).

the different test situations within each panel. This means that SOP and DOP can repeat their results from the laboratory situation at home testing stations.

4.1.1 Swiss olive oil panel (SOP)

Looking at the attribute fruitiness, **Figure 1** shows the comparison of the two medians per sample of all seven extra virgin olive oils. The maximum difference found was 0.4 for sample 249 (**Figure 2**). So, one can say, that—for the attribute fruitiness—there is no significant difference between results coming from *in situ* testing compared to home testing stations (remote). Results from mixed-model ANOVA are shown in **Table 5**.

For bitterness, you can easily see in **Figure 3** that there were found as well similar medians for all analyzed oils. The maximum difference was 0.3 for sample 104. Similar to fruitiness, as well as bitterness, there is no significant difference between results coming from the *in situ* testing compared to results from home testing stations (remote). Results from mixed-model ANOVA can be seen in **Table 6**.

For pungency, not surprisingly the comparison of medians in all seven oils in **Figure 4** shows only slight differences. The maximum difference found is 0.5 for sample 249. Again, there is no significant difference between results coming from the lab (*in situ*) compared to the home testing stations (remote). Results from mixed-model ANOVA can be seen in **Table 7**.

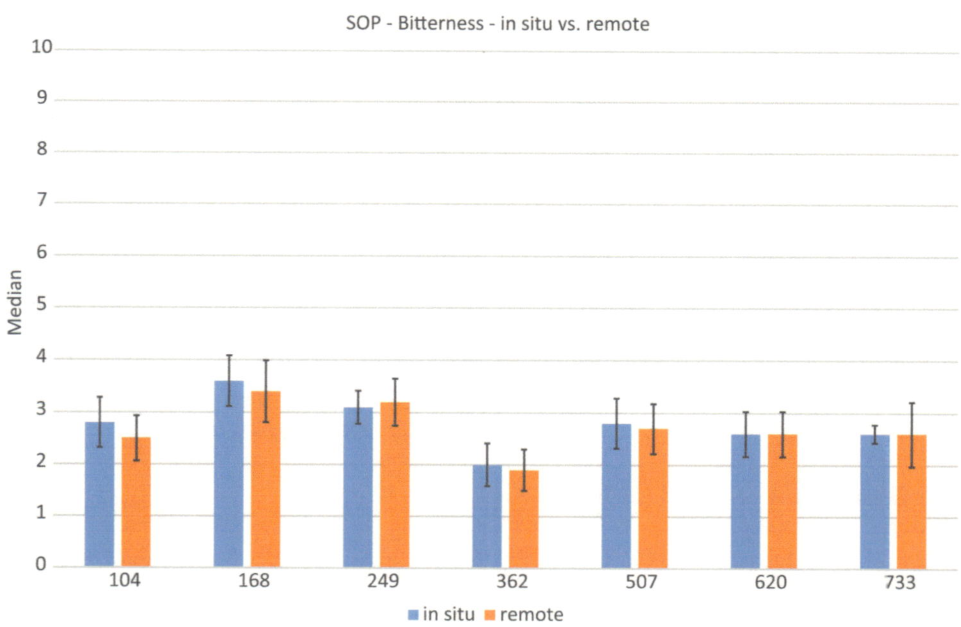

Figure 2.
Median of bitterness (SOP)—in situ versus remote (n = 9).

Contrast	Difference	Standardized difference	Critical value	P-value	Significant
In situ versus remote	0.142	0.991	2.120	0.336	No

Table 5.
Mixed model ANOVA for fruitiness (SOP)—in situ versus remote (n = 9).

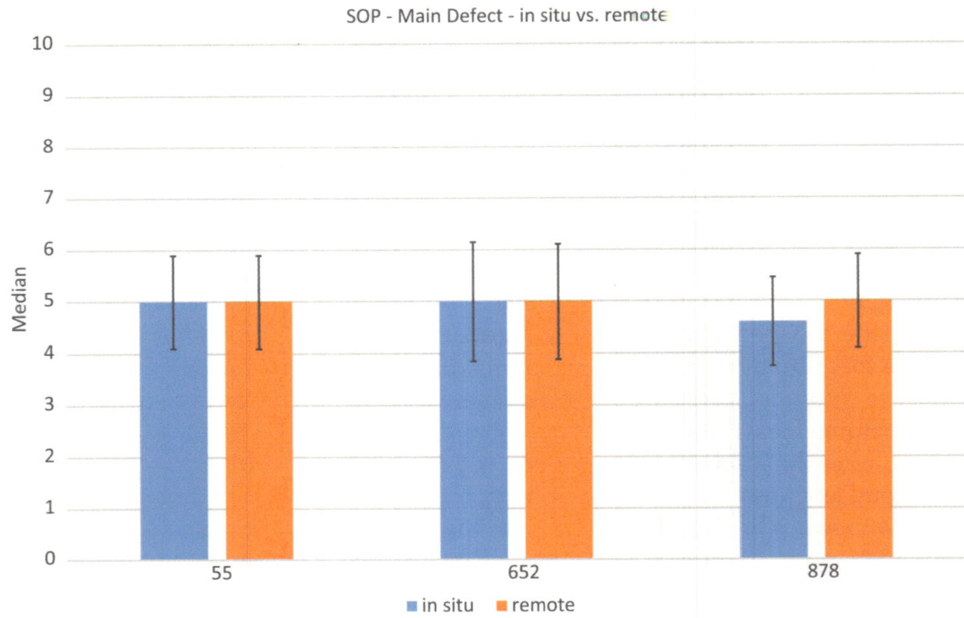

Figure 3.
Median of main defect (SOP)—in situ versus remote (n = 9).

Contrast	Difference	Standardized difference	Critical value	P-value	Significant
In situ versus remote	0.065	0.532	2.120	0.602	No

Table 6.
Mixed-model ANOVA for bitterness (SOP)—in situ versus remote (n = 9).

Finally, **Figure 5** focuses on defects and visualizes the comparison of medians of the main defect of the three defective olive oils. The maximum difference between medians was 0.4 for sample 878. So as well for defects, we can see no significant difference between results coming from the *in situ* testing compared to the home testing stations (remote). Results from mixed-model ANOVA are shown in **Table 8.**

4.1.2 German olive oil panel (DOP)

Looking at the attribute fruitiness, **Figure 6** visualizes the comparison of the medians of all seven extra virgin olive oils analyzed. Like for SOP, as well for DOP the maximum difference between medians for very small, in this case, 0.4 for sample 362. This proves that there is no significant difference between results coming from the *in situ* testing compared to the home testing stations (remote), as well for DOP. Results from mixed-model ANOVA are shown in **Table 9.**

For bitterness, **Figure 5** visualizes the comparison of the median of all extra virgin olive oils analyzed. The maximum difference between medians is 0.2 for samples 104 and 507. There is no significant difference between results coming from *in situ* testing compared to the home testing stations (remote). Results from mixed-model ANOVA are shown in **Table 10.**

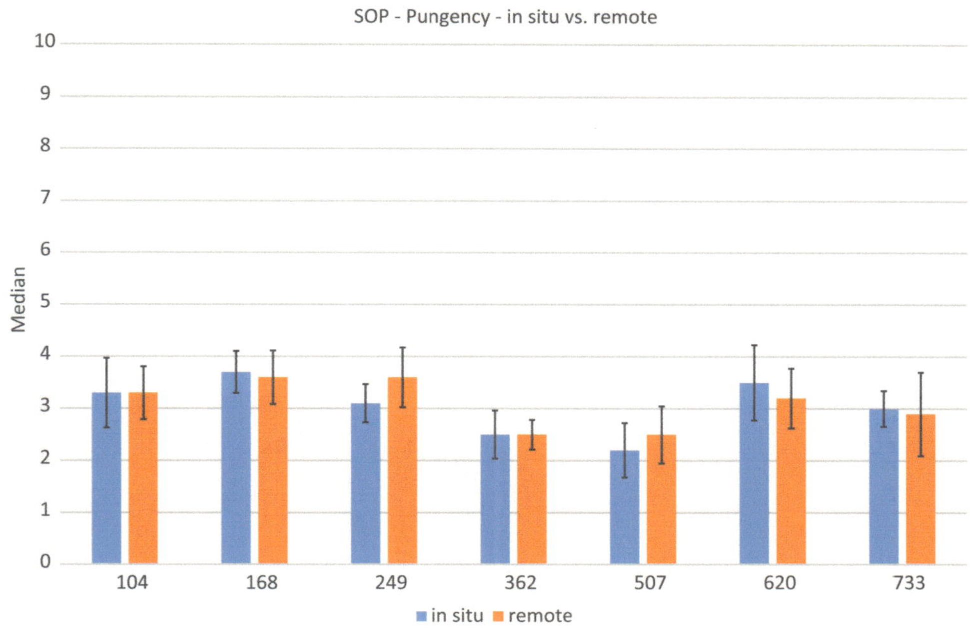

Figure 4.
Median of pungency (SOP)—in situ versus remote (n = 9).

Contrast	Difference	Standardized difference	Critical value	*P*-value	Significant
In situ versus remote	0.017	0.125	2.120	0.902	No

Table 7.
Mixed-model ANOVA for pungency (SOP)—in situ versus remote (n = 9).

For pungency, **Figure 7** visualizes the comparison of the median of all extra virgin olive oils analyzed. The maximum difference between medians is 0.3 for samples 104 and 362. There is no significant difference between results coming from the *in situ* testing compared to the home testing stations (remote). Results from mixed-model ANOVA are shown in **Table 11**.

Finally, **Figure 8** focuses on defects and visualizes the comparison of the median of the main defect of the 3 defective olive oils. The maximum difference between medians is 0.2 for samples 055 and 652. There is no significant difference between results coming from the *in situ* testing compared to the home testing stations (remote). Results from mixed-model ANOVA are shown in **Table 12**.

4.2 Agreement between different panels

Second, mixed-model ANOVA was used to analyze whether there exist differences between the data collected from the two IOC-recognized panels separately. It was found that there are statistically significant differences between the two panels. The largest difference for the mean value is 0.5 on a 10 cm scale. This means, that panels

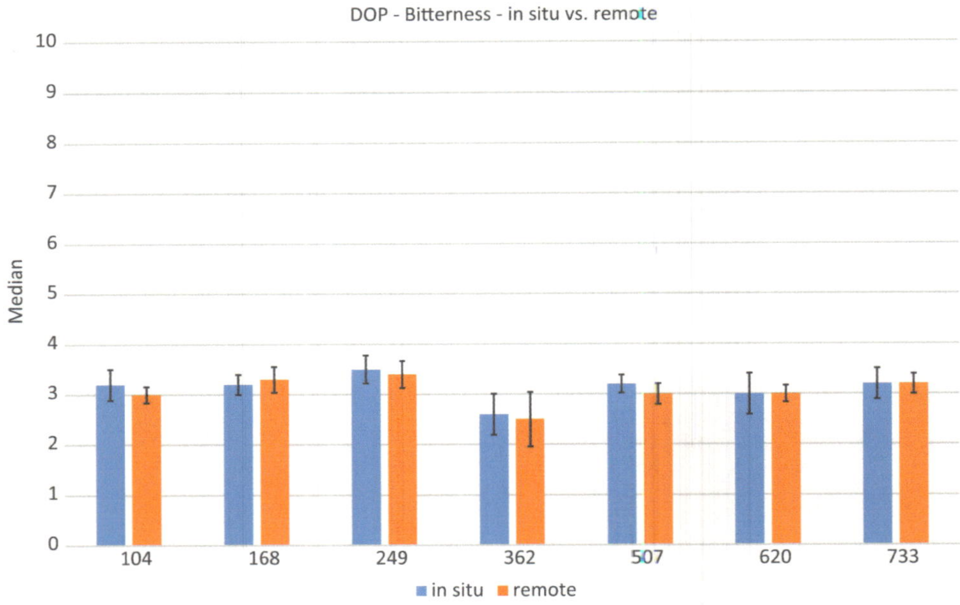

Figure 5.
Median of bitterness (DOP)—in situ versus remote (n = 11).

Contrast	Difference	Standardized difference	Critical value	*P*-value	Significant
In situ versus remote	0.082	0.289	2.086	0.775	No

Table 8.
Mixed-model ANOVA for main defects (SOP)—in situ versus remote (n = 9).

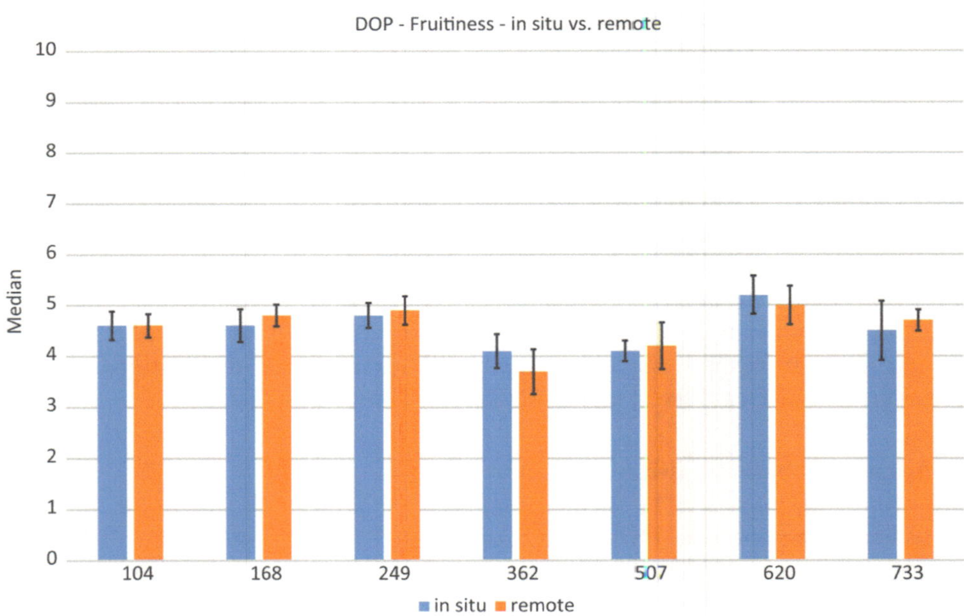

Figure 6.
Median of fruitiness (DOP)—in situ versus remote (n = 11).

Contrast	Difference	Standardized difference	Critical value	*P*-value	Significant
In situ versus remote	0.004	0.044	2.086	0.966	No

Table 9.
Mixed-model ANOVA for fruitiness (DOP)—in situ versus remote (n = 11).

Contrast	Difference	Standardized difference	Critical value	*P*-value	Significant
In situ versus remote	0.095	1.500	2.086	0.149	No

Table 10.
Mixed-model ANOVA for bitterness (DOP)—in situ versus remote (n = 11).

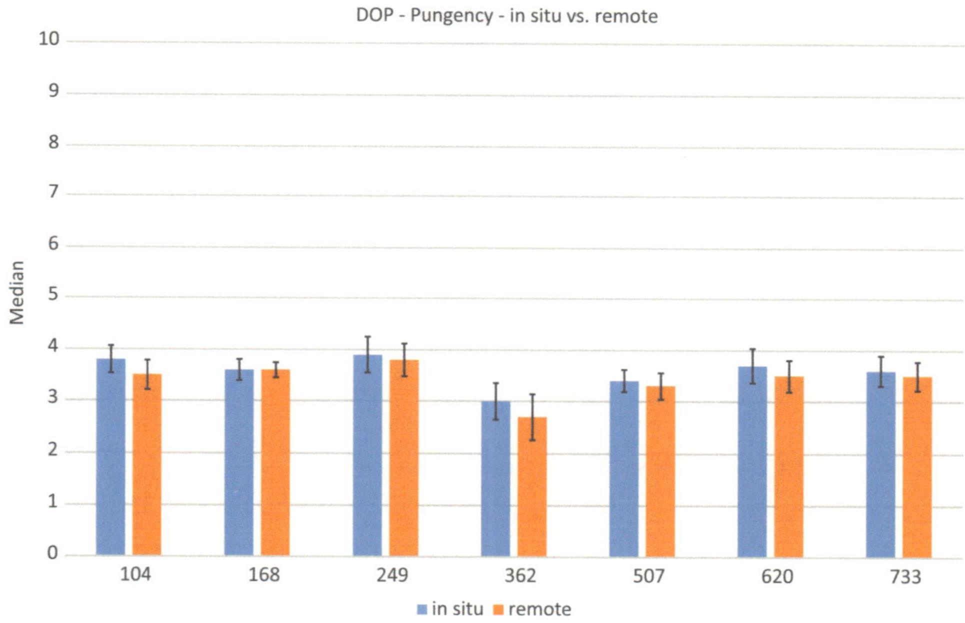

Figure 7.
Median of pungency (DOP)—in situ versus remote (n = 11).

Contrast	Difference	Standardized difference	Critical value	*P*-value	Significant
In situ versus remote	0.134	1.958	2.086	0.064	No

Table 11.
Mixed-model ANOVA for pungency (DOP)—in situ versus remote (n = 11).

show variance, but nevertheless, results are within the expected and accepted variation proposed by the IOC.

For the attribute fruitiness "*in situ*" as well as fruitiness "remote" **Figures** 9 and 10 show that the two panels do differ only slightly (0.2) and the results of the two panels show significant differences (**Table 13**).

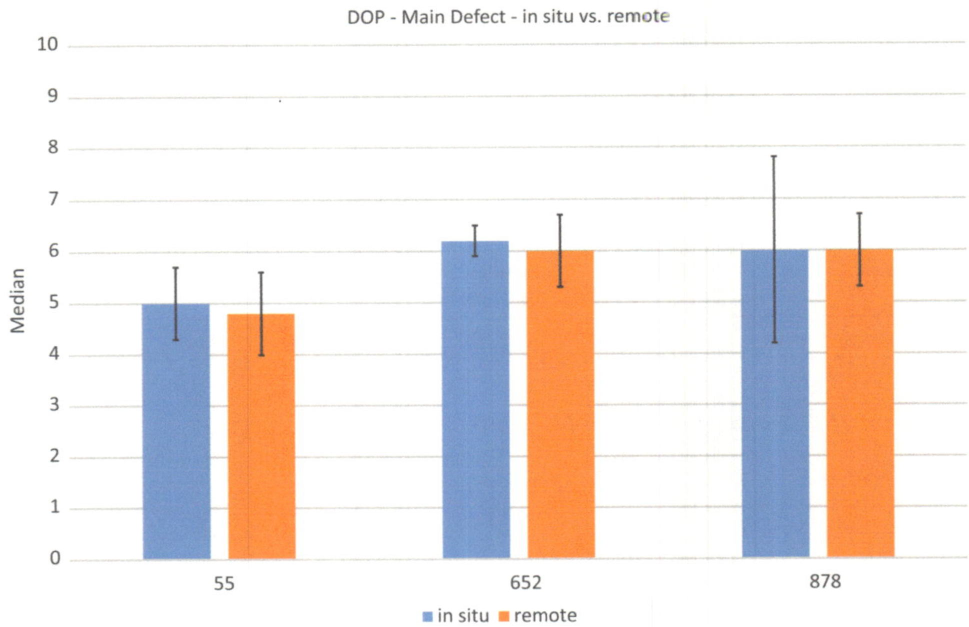

Figure 8.
Median of main defect (DOP)—in situ versus remote (n = 11).

Contrast	Difference	Standardized difference	Critical value	*P*-value	Significant
In situ versus remote	0.078	0.142	2.120	0.889	No

Table 12.
Mixed-model ANOVA for main defects (DOP)—in situ versus remote (n = 11).

This is as well the case for the other positive attributes—bitterness and pungency. Regardless that the comparisons of panels show significant differences, it can be stated, that the maximum deviation in the mean value was in all attributes only 0.5.

A similar situation is found for the main defects "*in situ*" and "remote." It is shown in **Figures 11** and **12** that the two panels do differ slightly (1.5) and results from both panels show significant differences (**Table 14**).

Based on the shown data, it can be concluded, that even if the "difference between panels" (DOP/SOP) is significant, the variance in all cases is well below the IOC accepted differences between recognized panels.

4.3 Panel performance of single panels

Third, and based on the document COI/T.28/Doc. No.1/Rev. 52,019, the panel performance for both panels were analyzed according to the following selected criteria: *Z*-Score, Deviation Number as well as Normalized Error and Precision Number for both participating panels, SOP and DOP. Results show that both panels meet the IOC requirements in all criteria mentioned. The following data from SOP are shown exemplarily.

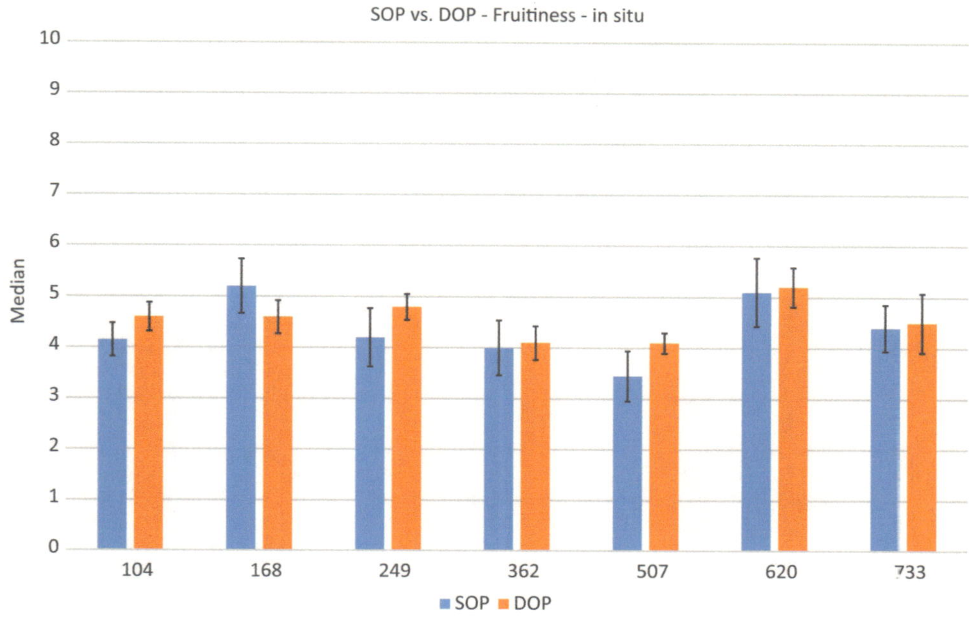

Figure 9.
Median of fruitiness (SOP versus DOP)—in situ.

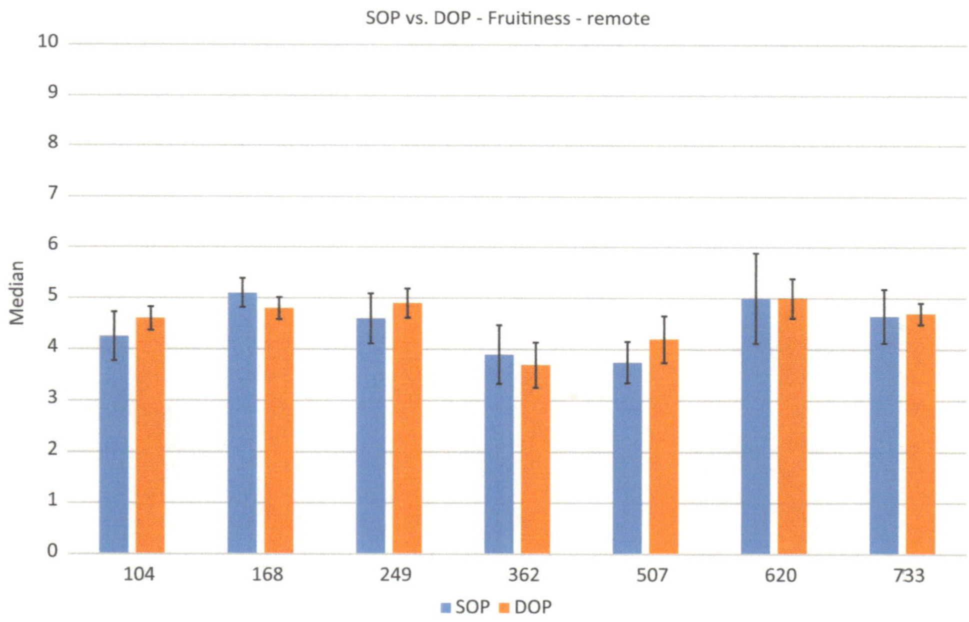

Figure 10.
Median of fruitiness (SOP versus DOP)—remote.

Contrast	Difference	Standardized difference	Critical value	*P*-value	Significant
SOP versus DOP	0.181	2.702	1.978	0.008	Yes

Table 13.
Mixed-model ANOVA for fruitiness (SOP versus DOP).

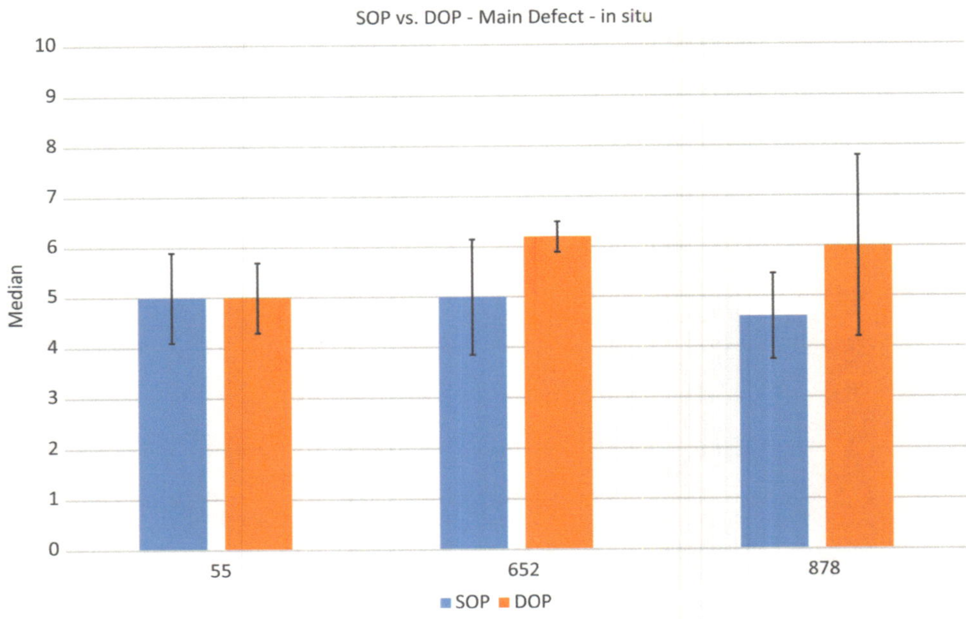

Figure 11.
Median of main defect (SOP versus DOP)—in situ.

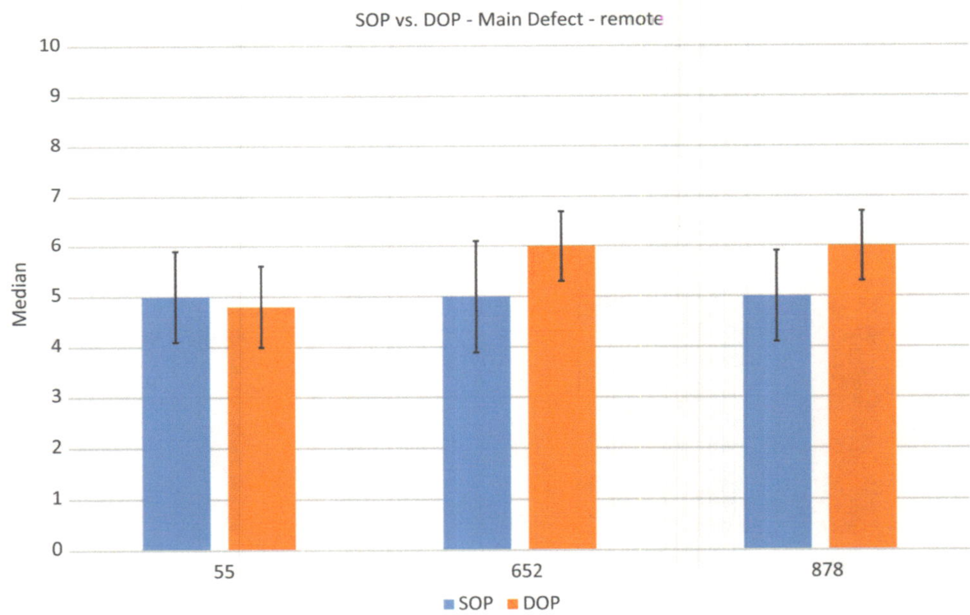

Figure 12.
Median of main defect (SOP versus DOP)—remote.

4.3.1 Trueness (homogeneity): z-score (for panels)

The calculation of the z-score for panels focuses on the difference between one-panel result (median) and a reference result (median) in relation to a defined SD of

DOI: http://dx.doi.org/10.5772/intechopen.102702

Contrast	Difference	Standardized difference	Critical value	*P*-value	Significant
SOP versus DOP	1.449	3.853	2.002	<0.0001	Yes

Table 14.
Mixed-model ANOVA for main defect (SOP versus DOP).

0.7. The reference result (in this study) is defined as the mean over all four test results (DOP remote, DOP *in situ*, SOP remote, and SOP *in situ*).

Z-score = difference between one-panel result (median) and the reference result* (median) in relation to SD.

*Reference result = median of results from all four considered test situations (DOP remote, DOP *in situ*, SOP remote, and SOP *in situ*)

- SD = maximum standard deviation of the method = ±0.7
- Warning limit = ±2
- Action limit = ±3

→ Proof of trueness/homogeneous results (statistically acceptable)

We can see in **Figures 13** and **14** for the attribute fruitiness, that for *in situ* testing as well as for remote testing, the *z*-Score of SOP is in line with the requirements, that is—well below warning (±2) and action limit (±3). This is the case for bitterness and pungency as well, but not shown here. Based on these findings, it can be concluded that results from SOP are homogeneous (in both test situations) and statistically acceptable (= aspect of trueness).

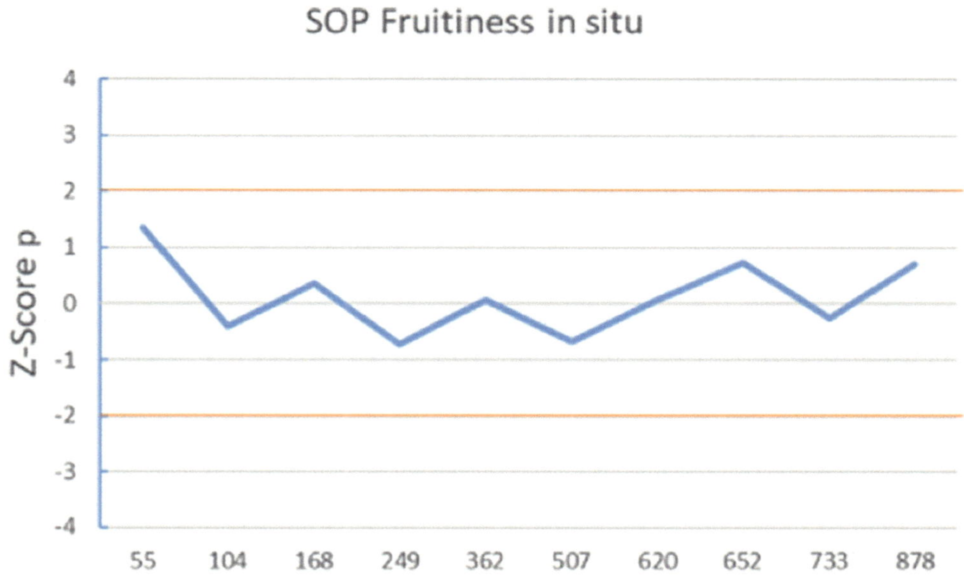

Figure 13.
z-Score (SOP) fruitiness—in situ (n = 9).

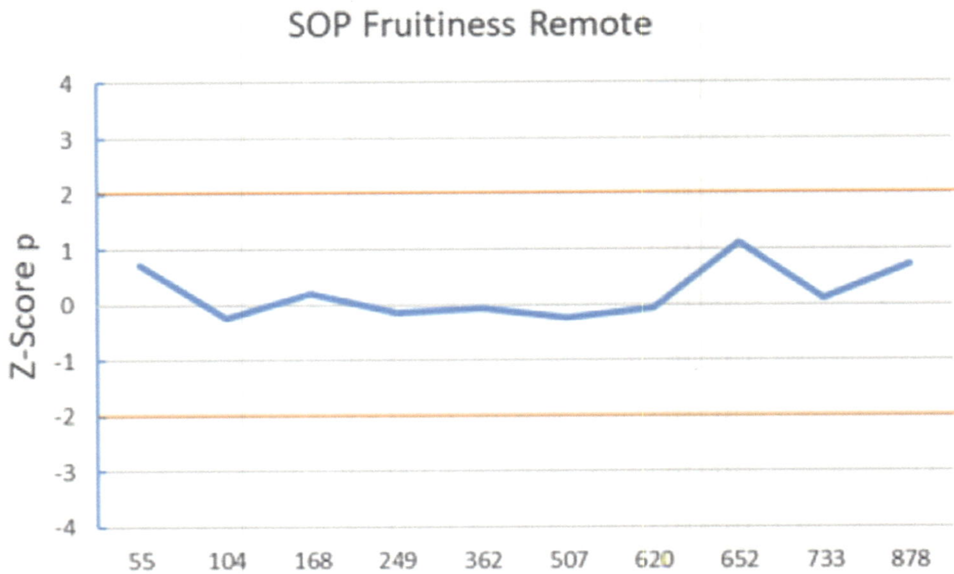

Figure 14.
z-Score (SOP) fruitiness—remote (n = 9).

4.3.2 Trueness (homogeneity): DN_p *(deviation number for panels)*

The calculation of the DN_p (deviation number for panels) focuses on the sum of differences (squared) between duplicate results (median) and the reference result (median) in relation to the number of reference samples (in our case 4). The reference result (in this study) is defined as the mean over all four test results.

DN_p = sum of differences (squared) between duplicate results (median) and the reference result* (median) in relation to the number of reference samples (4).

*Reference result = median of results from all considered test situations (DOP remote, DOP *in situ*, SOP remote, and SOP *in situ*)

- Duplicate = comparison between *in situ*/remote

- Number of samples building the reference mean = 4

→ Proof of trueness/homogeneous results (statistically acceptable)

Figure 15 shows for the attribute fruitiness, that the deviation number for SOP is in line with the requirements, that is: well below the warning limit of 1.0 and the action limit of 2.0). This is the case for bitterness and pungency as well, not shown here. Based on these findings, it is proven that results of SOP are homogeneous between different test situations and statistically acceptable (= aspect of trueness).

4.3.3 Precision (repeatability, consistency): E_n *(normalized error)*

The calculation of the E_n (normalized error) focuses on the difference between duplicate panel results (mean) in relation to a defined SD of 0.7.

E_n = difference between duplicate panel results (mean) in relation to SD.

- Duplicate = comparison between *in situ*/remote

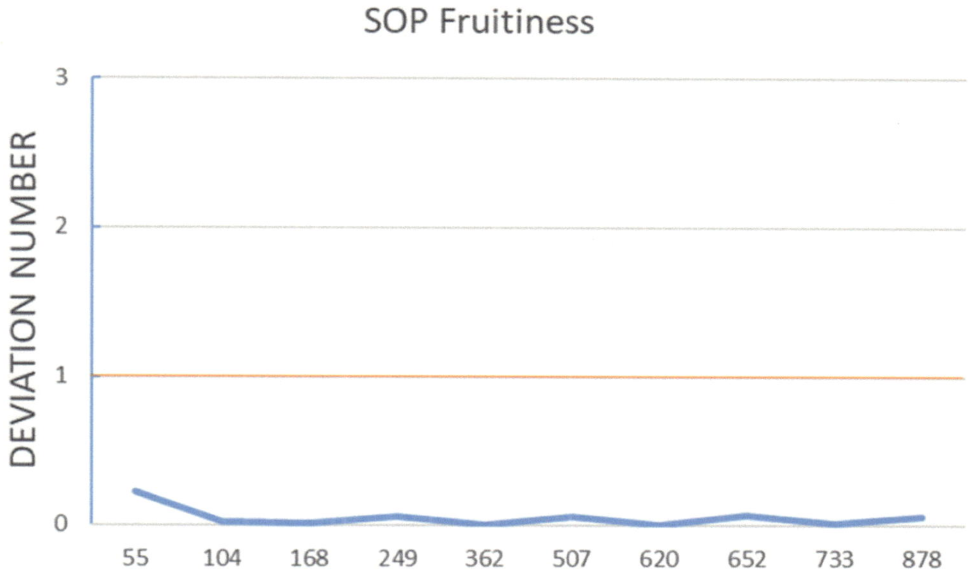

Figure 15.
DN_p (SOP) fruitiness (n = 9).

- SD = maximum standard deviation of the method (or maximum error) ± 0.7

→ Proof of precision/consistent results (statistically acceptable)

Coming to the aspect of "Precision" (repeatability, consistency), we can see in the figure for fruitiness (**Figure 16**) that results of the normalized error from SOP are in line with the requirements, that is—well below the action limit of 1.0. This is the case for bitterness and pungency as well, not shown here. Based on these findings, it can be concluded that results of SOP are repeatable between different test situations and statistically acceptable (precise, consistent).

4.3.4 Precision (repeatability, consistency): PNp (the precision number for panels)

The calculation of the PN_p (the precision number for panels) focuses on the sum of differences (squared) between duplicate panel results (mean) in relation to the number of duplicate samples (in this study 10).

PN_p = sum of differences (squared) between duplicate panel results (mean) in relation to the number of duplicate samples (10)

- Duplicate = comparison between *in situ*/remote
- Number of duplicate samples = 10

→ Proof of precision/consistent results (statistically acceptable)

We can see in the figure for fruitiness (**Figure 17**) that the results of the precision number for panels from SOP are in line with the requirements, that is—well below the action limit of 2.0. This is the case for bitterness and pungency as well, but not shown here. Based on these findings, it is proven that results of SOP are consistent and precise between different test situations and statistically acceptable.

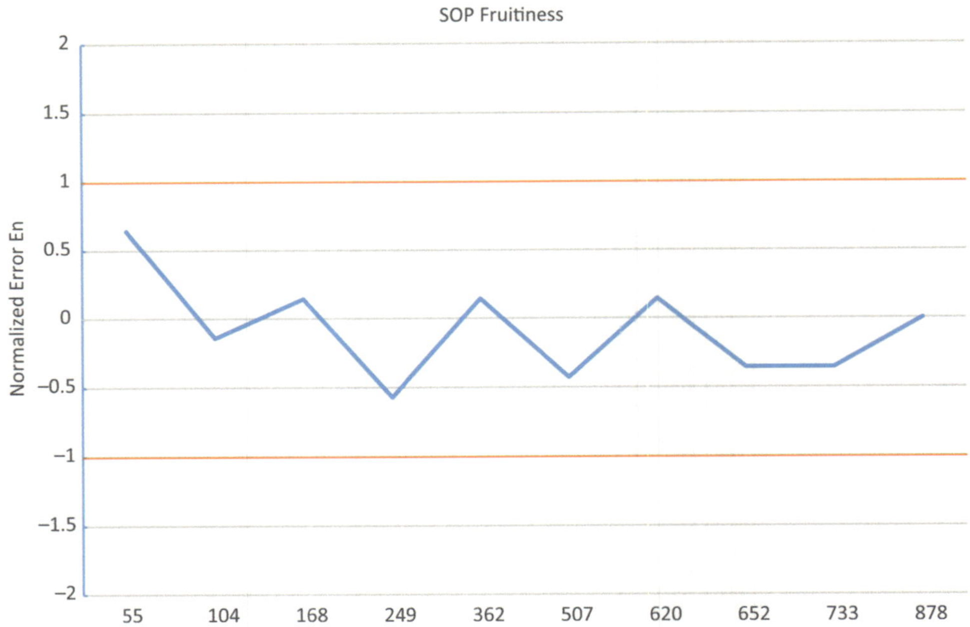

Figure 16.
E_n (SOP) fruitiness (n = 9).

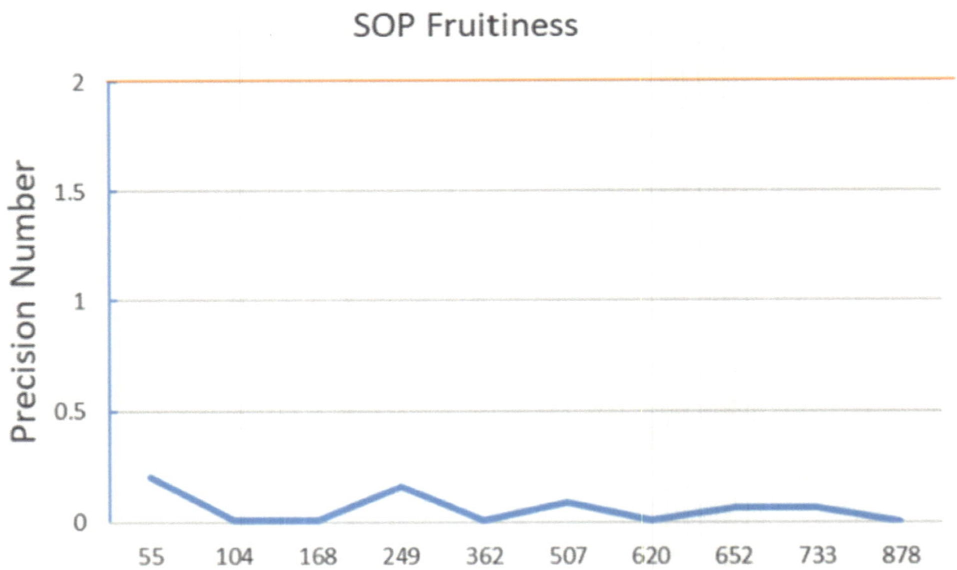

Figure 17.
PN_p (SOP) fruitiness (n = 9).

5. Conclusions and outlook

To prove data quality in terms of reliability and validity, 10 olive oils (seven of them extra virgin and three defective ones) were independently tested by the Swiss Olive Oil Panel/SOP (nine panelists) and the German Olive Oil Panel/DOP

(11 panelists) in different test situations—namely in sensory laboratories (*in situ*) and at so-called home testing stations.

Analyzing the raw data, various aspects of panel performance were looked at—especially the different test situations (*in situ* versus remote), the variations between results from different panels as well as selected aspects concerning the individual panel performance per panel.

1. The individual panel performance per panel [3] shows that both panels meet the requirements from IOC.

2. Significant differences between the two considered olive oil panels were found, but the variation is within the accepted limits required by the IOC.

3. No significant differences between test situations (*in situ*/remote) were found → panels can repeat results between laboratory and home testing situation.

Overall, results show that the influence of testing through different panels (SOP versus DOP) is bigger than the impact of testing in different test situations (*in situ* versus remote). However, it is important to mention that to achieve such reliable data in both test situations, professional equipment and intense and adequate training of the panelists/panels is required. DOP and SOP fully meet these expectations. Data from both panels coming from both test situations are fully reliable and valid.

Besides the convincing findings from this study—many comparison tests over the last years took place proofing reliability and precision of tests taking place either *in situ* or remote. Remote testing—in a defined framework—therefore is proven to be a very valuable methodology and technique, independently from pandemic situations. In the meantime, the test procedure is well established in different contexts and accepted by official certification bodies (ISO 17025) as well as retailers, importers, and consumers.

In the future, it will be valuable to set up advanced follow-up studies with even more participating panels from different countries to regularly confirm findings and strengthen the trust in the data and conclusions of the study at hand.

Acknowledgements

Our thanks go especially to the panelists from the two involved panels—the German Olive Oil Panel DOP and the Swiss Olive Oil Panel SOP—as well as to their panel leaders: Richard Retsch from DOP and Martin Popp/Annette Bongartz from SOP.

Conflict of interest

The authors declare no conflict of interest.

Author details

Annette Bongartz[1]*, Martin Popp[1] and Richard Retsch[2]

1 Zurich University of Applied Sciences (ZHAW)/Swiss Olive Oil Panel (SOP), Wädenswil, Switzerland

2 German Olive Oil Panel (DOP) e.V., Nuremberg, Germany

*Address all correspondence to: annette.bongartz@zhaw.ch

References

[1] Commission Regulation (EEC) No 2568/91 of 11 July 1991 on the characteristics of olive oil and olive-residue oil and on the relevant methods of analysis. 1991 [as amended]

[2] COI/T20/Doc. No 15/Rev. 10 2018. Sensory analysis of olive oil—Method for the organoleptic assessment of virgin olive oil. 2018

[3] COI/T28/Doc. No 1/Rev. 5 2019. Guidelines for the accomplishment of requirements of standard ISO 17025 of testing laboratories with particular reference to virgin olive oil. 2019

[4] Decision No Dec-III.1/113-VI/2020. Adopting the guidlines for the managament of virgin olive oil tasting panels in the event of a pandemic. 2020

[5] EN ISO/IEC 17025:2017. General requirements for the competence of testing and calibration laboratories. 2017

[6] Dinnella C, Pierguidi L, Spinelli S, Borgogno M, Toschi TG, Predieri S, et al. Remote testing: Sensory test during Covid-19 pandemic and beyond. Food Quality and Preference. 2022;**96**:104437. DOI: 10.1016/j.foodqual.2021.104437

Chapter 6

Table Olives: Toward Mechanical Harvesting

Iris Biton, Dvora Namdar, Yair Mani and Giora Ben Ari

Abstract

The major reasons for developing mechanical technologies for olive harvesting are the chronic shortage of workers for manual harvesting and increasing labor costs. To enable these technologies to operate, new table olive cultivars suitable for mechanical harvesting are necessary. The two major factors required for the shift from manual to mechanical harvesting of table olives are improved harvesting efficiency and prevention of fruit injury. Improved harvesting efficiency requires suitable pretreatment to enable fruit abscission with minimal defoliation, even when the harvesting is performed by a trunk shaker. The second requirement is prevention of external fruit color change or browning as a result of fruit injury, by development of olive cultivars with firm skin and higher resistance to the bruising caused by mechanical harvesting. This genetic adaptation to mechanical harvesting must be accompanied by efficient post-harvest processing of the olives. In this chapter, we will review the published studies regarding mechanical harvest of table olives, and attempt to identify the main issues, which still need to be studied in order to facilitate the transition from hand to mechanical harvest of table olives.

Keywords: table olive, mechanical harvest, olive cultivars, fruit injury, trunk shaker, harvester

1. Introduction

Table olives (*Olea europaea* L.) have been consumed by populations surrounding the Mediterranean basin, as long as 7000 years ago [1]. Their consumption has expanded to many other countries due to the increasing popularity of the Mediterranean diet [2]. The World Catalog of Olive Cultivars reports about 2500 olive varieties, and their selected use–for oil, table or both, is determined by given parameters [3]. Of these, table olives account for an annual worldwide production of close to 3 million tons.

In general, varieties of table olives are mostly low in oil content, medium to large in size, with flesh-to-pit ratios higher than 4:1 and appropriate texture as set by the International Olive Oil Council (IOC) [4]. The olive trees produce drupes consisting of an epidermis and a soft mesocarp surrounding a stone containing the seed [5]. The epidermis (1.5–3% of the total weight), is constituted mainly of cellulose and cutin, its main function being to protect against external infestations or injury [6]. The olive mesocarp constitutes 70–90% of the fruit weight, the stone accounts for another 5–30%, and the seed is about 1–3% of the fruit, by weight [7].

The five major producers of table olives today are (listed in alphabetic order) Algeria, Egypt, Greece, Spain and Turkey [4]. According to the IOC 2019 report, world table olive production exceeded 2.9 million tons for the season of 2017/2018 [8]. Production is increasing in other regions, such as South America, Australia, and the Middle East with Italy and Portugal also being major producers [9]. Israel has also developed several varieties of table olives, such as 'Kadeshon' and 'Lavee' [10]. However, these varieties play a minor role in the world market.

Classification of table olives is based on standards set by the International Olive Council in 2008 and reaffirmed in 2011 (COI/OT/MO/Doc. No 1. Method for the sensory analysis of table olives and Decision No DEC-18/99-V/2011 COI/OT/MO No 1/Rev. 2). Classification of table olives is based on the median Defect Predominantly Perceived (DPP). Four categories were designated: (1) Extra or Fancy: DPP ≤ 3; (2) First Choice or Select: $3 < \text{DPP} \leq 4.5$; (3) Second or Standard: $4.5 < \text{DPP} \leq 7.0$; (4) Olives that may not be sold as table olives: DPP > 7.0. Hardness, crunchiness and fibrousness have also been characterized in table olives [11]. Thus, it is important to be aware of the external characteristics of table olives, such as texture and appearance, their importance in the demand for table olives and their role in determining the price to be obtained in the marketplace.

For these reasons, table olives are traditionally harvested manually. It is interesting to note that despite being one of the oldest domesticated fruit crops in the world [1], table olives have benefitted from few technological innovations, especially in regard to harvesting [12]. Because their external appearance and texture are so important in their marketing appeal, special caution must be practiced at harvest. However, a chronic shortage of workers and high labor costs have prompted the need for other, more efficient and economic solutions. Mechanical harvesting methods practiced today are (1) trunk shaking, which can be applied simultaneously with rod beating; and (2) use of an overhead harvester. To promote harvest efficiency, application of an abscission agent should also be considered. Different cultivars may react differently to these mechanical and agronomical methods. In their research, Zipori et al. [13] compared the harvesting efficiency and final product quality of four cultivars of green table olive, 'Manzanillo', 'Hojiblanca', 'Souri', and 'Nabali Mouhassan', in response to manual versus trunk shaker harvesting.

When mechanical harvesting was supplemented by rod beating, harvesting efficiency reached 80–95%, similar to the efficiency of manual picking. However, harvesting efficiency of the trunk shaker without rod beating was significantly lower [13]. Interestingly, application of an abscission agent to accelerate fruit detachment and facilitate mechanical harvest, did not improve harvesting efficiency. Cultivar dependence was also observed: 'Hojiblanca', 'Souri' and 'Nabali Mouhassan' varieties showed similar final product quality using either method, while the quality of 'Manzanillo' olives harvested mechanically was inferior to those harvested manually.

The high sensitivity of table olives to damage caused by mechanical harvesting limits the suitability of this rather efficient method. An objective determination and quantification of the damage caused to table olives by mechanical harvesting may be obtained by digital image analysis. Investigation of fruit damage conducted on 'Manzanillo' table olives in Seville, Spain showed that mechanical harvesting with a trunk shaker led to a rate of bruising 12 times greater than that obtained by manual harvesting. About 60% of the damage to fruit was attributed to fruit-fruit abrasion, fruit-branch contact, and friction from the vibration of the fruit in the tree canopy during harvesting [14]. Most external bruising appeared within the first hour after harvesting.

Another recently developed harvesting technology is the New Pneumatic Harvester (NPH). Two oil cultivars ('Mari' and 'Yellow') were selected to evaluate the NPH system. Harvesting capacity and efficiency, leaf loss and fruit damage were measured. Results showed that the NPH harvested 92% of olive fruits. The percent of leaf loss during the harvesting process was 2.55%. The collector system reduced the level of damaged fruit from 61–25% in both tested cultivars [15].

The mean value of harvesting efficiency with trunk shakers is 72–74%, when applied without the addition of rod beating or abscission agents [16, 17]. In order to achieve harvesting efficiency greater than 85%, tree trunk vibration parameters were set above an acceleration value of 183.4 m/s^2, and at a frequency of 28.1 Hz. Although increasing the trunk acceleration improved harvesting efficiency, it led to an increase in damage to the harvested fruit 3.5 times greater than the damage caused by manual harvesting [16]. This technique also caused damage to the tree trunk [17].

Olive orchards should be specially prepared for this technology before its application. Ferguson and Garcia [12] studied the effects of pruning on fruit yield that was mechanically harvested. They compared the yield achieved by mechanical harvesting after two different pruning methods - mechanical pruning during six consequential years to that of manual pruning. They reported that use of mechanical pruning resulted in a harvest of 92% of the total yield on the trees where only 81% was harvested from trees that had been hand-pruned. There were no significant differences in the percentage of fruit yield and fruit size as a result of the different pruning treatments. They concluded that the use of mechanical pruning does not decrease average annual yields. These results suggest that in addition to the use of mechanical harvesting, pruning can also be done mechanically without lowering the yield of the tree, thus reducing the management costs of the trees "trained" and adapted for.

Mechanical harvesting can also be carried out by use of an overhead grape harvester. To use this device, the olive grove must be planted with suitable varieties with rows aligned specially for the harvester. New grove designs and management practices such as super-high-density groves which are used in oil olive production should be developed as an option for mechanical harvesting of table olives. In 2012, two table olive cultivars, 'Manzanillo de Sevilla' and 'Manzanillo Cacereña', were harvested in a 5-year-old super-high-density grove (1975 trees/ha) after being planted in continuous hedgerows (≈10,000 and 18,000 $kg{\cdot}ha^{-1}$, respectively). The differences between manual and mechanical harvesting (using a grape straddle harvester), in time, efficiency, and fruit quality were assessed [18]. The average harvest time per hectare with a grape straddle harvester was less than 1.7 hours compared with minimum of 576 man-hours/hectare for manual harvest. Fruit removal efficiency was high in both cases (98%). The mechanically harvested olives had a very high rate of bruised fruits (>90%). The severity of the damage was greater in 'Manzanillo de Sevilla' than in 'Manzanillo Cacereña'. After Spanish-style green processing, however, the proportion of bruised fruits was below 3% for each cultivar. Mechanically harvested fruits showed a significantly higher proportion of cutting (18%), a type of damage that may take place during harvesting, and reduced firmness than those harvested manually [18].

2. The challenge

It is clear that growers, technologists and scientists must join in a partnership developing and screening cultivars suitable for mechanical harvesting, either by overhead harvester or trunk shaker. Developing varieties with harder or thicker pills may help in achieving this goal. Another approach would be to investigate the mechanism of fruit detachment in olives, and develop a selective treatment which is commercially practical, and does not cause leaf abscission. A possible reason for the long survival of traditional harvest methods for table olives may be the low efficiency of the detachment mechanism in olive fruits. Multiple analyses carried out by the authors show that mature olive fruits do not produce or release ethylene to any detectable level. It seems therefore, that the fruit detachment mechanism in olives is not regulated by ethylene release, but by a different mechanism.

The desired characteristics of olives fit for table use include larger fruit size, firmness of the flesh of the fruit, resistance to disease, high flesh-to-seed ratio; skin thickness desirable for eating, resistance of epicarp to cracks and bruising and the ability to maintain intact appearance.

One of the main challenges facing growers of table olives interested in converting to mechanical harvesting, is reducing damage to the fruit caused by the harvesting process. This may be achieved by combining the two parameters: (1) enhancing pill durability by increasing thickness of the pill [19]. As the pill thickens, the fruit is more resistant to damage. However, this must be balanced against sensory sensitivity to the feel of the olive in the mouth (2) Reduction of the attachment force of the fruit to the branch, without increasing leaf abscission.

In developing strategies to reduce bruising of the olives at harvest, it is important to understand major factors influencing bruise susceptibility of fresh fruits. Excessive compression forces and a series of impacts during harvesting can cause severe bruise damage. In addition to mechanical forces applied to the fruit and the tree, the stage of fruit maturity also affects bruising. (The susceptibility to bruising depends partly on physiological and biochemical variables. Environmental factors such as temperature, humidity and post-harvest treatments may play a role in susceptibility to fruit damage [20]. Thus, mechanical harvesting must be performed properly in order to reduce both frequency and severity of bruising and increase the resistance of fresh fruit to bruise damage.

As olive varieties differ in many qualities such as heat and disease resistance, fruit size, yield and many more traits [10, 21–23], they differ also in their resistance to shock and bruising. Jiménez et al. [24] observed significant differences among 14 selected genotypes in sensitivity to bruising. Histological sections of bruised and unaffected fruit tissues revealed a subsurface zone of tissue discoloration but a much greater area in which cell structure was disrupted. To further assess the susceptibility of different table olives varieties to bruising, Jiménez et al. [25] studied damages incurred by two different cultivars–'Manzanillo de Sevilla' and 'Hojiblanca', at 4 and 24 h after impact induced bruising. The predominant post bruising changes they noted in the mesocarp included ruptured cells, cell wall thinning, and discoloration. These changes appeared greater in 'Manzanillo de Sevilla' than 'Hojiblanca' and were more evident 24 h after the impact. This verifies the assumption that different table olive cultivars may demonstrate differential resistance to bruising. Furthermore, the authors noted several factors that may serve as parameters defining the level of damage: total damaged area, the number of tissue ruptures in the mesocarp, and the distance from the fruit exterior to the nearest tissue rupture were different in the two cultivars. These factors were recommended by the authors as useful parameters characterizing susceptibility to bruising among table olives.

Another criterion useful in assessing bruised fruit is the proportional area of brown coloring after injury, compared to the total fruit surface area. Using the knowledge that mesocarp cells are damaged when pressure is applied, Goldental-Cohen et al. [19] screened 106 olive cultivars of the Israeli germplasm collection managed at the Volcani Institute for sensitivity to browning in response to injury. Using the above criterion, they showed that post-bruise browning may vary from 0 to 83.61%. Fourteen genetically different cultivars did not brown 3 h after application of pressure. Among them, some cultivars may be selected as suitable table olives. Cultivars resistant to browning were found to have thicker cuticles than those of susceptible varieties, thus cuticle thickness may very well be an important parameter in selecting table olive cultivars suitable for mechanical harvest. A shift to browning-resistant cultivars in place of the popular cultivars currently in use will encourage mechanical harvest of table olive without affecting fruit quality.

Overcoming the bruising caused by mechanical harvest may be aided by adopting the proper irrigation regime. Casanova et al. [26] showed that fruits under regulated deficit irrigation were less susceptible to bruising than fruits fully irrigated. Fruits of 'Manzanillo de Sevilla' that were subjected to regulated deficit irrigation and full irrigation were bruise-induced by a standardized mechanical blow to evaluate bruising susceptibility. Damage was evaluated 3 h after treatment. Fruits under restricted water regimes in the weeks before harvest were much less susceptive to bruising [27]. Thus, controlling irrigation may overcome bruising effects in existing orchards, even for more susceptible varieties.

The low efficiency of hand harvesting of table olives and the rise in the costs of labor prioritizes the need for mechanical harvesting. To achieve mechanical harvesting, the attachment force of fruit to the branch should be lower than 200 gr. Today, most farmers use a pre-harvest abscission compound in order to decrease the fruit detachment force to a level less than 200 gr before mechanically harvesting their oil producing groves. However, table olives are harvested before ripening. At this early stage, the detachment force of the fruits is still very high and does not differ significantly from the detachment force of the leaves. A selective abscission compound is crucial for adapting table olive groves to mechanical harvesting. Goldental-Cohen et al. [28] studied the anatomical and molecular differences between the fruit and leaf abscission zones in olive. Typically, the abscission zone is characterized by small cells with less pectin compared to neighboring cells. This type of cell is found in the leaf but not in the fruit abscission zone. The fruit abscission zone 3 (FAZ3), located between the fruit and the pedicel, was found to be the active AZ in mature fruits. In an attempt to differentiate between fruit and leaf AZs, olive-bearing trees were treated with ethephon, an ethylene-releasing compound, and the effect of this treatment on the detachment force of fruits and leaves 5 days after its application was determined. Transcriptomic analysis of the various abscission zones revealed induction of genes involved in oxidation stress specifically in the leaf abscission zone. They found that adding antioxidants such as ascorbic acid or butyric acid to the ethephon inhibited leaf abscission but enhanced fruit abscission. Treating olive-bearing trees with a combination of ethephon and antioxidants reduces the detachment force of fruit without weakening that of the leaves. Hence, this selective abscission treatment may be used to promote mechanized harvest of olives [28].

Another way to reduce the consequences of bruising by mechanical harvesting is by application of post-harvest treatments. Zipori et al. [29] studied postharvest field treatments on 'Manzanillo' olives, the main table olive cultivar in Israel. This variety is highly sensitive to bruising and other damage caused by mechanical harvesting. Immersing the fruit in a 1% NaOH solution immediately after harvest seems to be the most effective treatment among those studied. This treatment reduced the percentage of severely bruised fruit to reasonable values. We suggest that this treatment, together with other advances in the field, such as improved shakers and\or use of abscission agents, will enable cost-effective mechanical harvesting of 'Manzanillo' table olives [29].

3. Conclusions and further perspectives

The increasing demand for healthy foods has stimulated the industry to study more intensively the existing varieties of table olives found in Mediterranean countries. Olives are considered by many in the food industry to be the "food of the future" [30]. Their balanced fatty acid content and the presence of significant concentrations of polyphenols and fibers increases their attractiveness to the modern consumer. The variety of methods and styles of preparation further increases their demand. For these reasons, most research during the last two decades has focused on the effects of pre-harvesting care and processing technologies, on the nutritional and sensory properties of the different varieties of table olives. Technical aspects of harvesting and post-harvesting processes seem to have been neglected, and to a large extent, traditional methods of handling the produce were retained.

The transition to mechanical harvesting of table olives is essential for the economic survival of this branch of agriculture just as it was for oil olives. The costs of manual labor are increasing, thus lowering profitability to a dangerous level. The changes suggested in this chapter, and the availability of the technologies enabling these changes, are necessary for adapting local, family-sized olive orchards to the scale needed to meet the demands of the world market of the future.

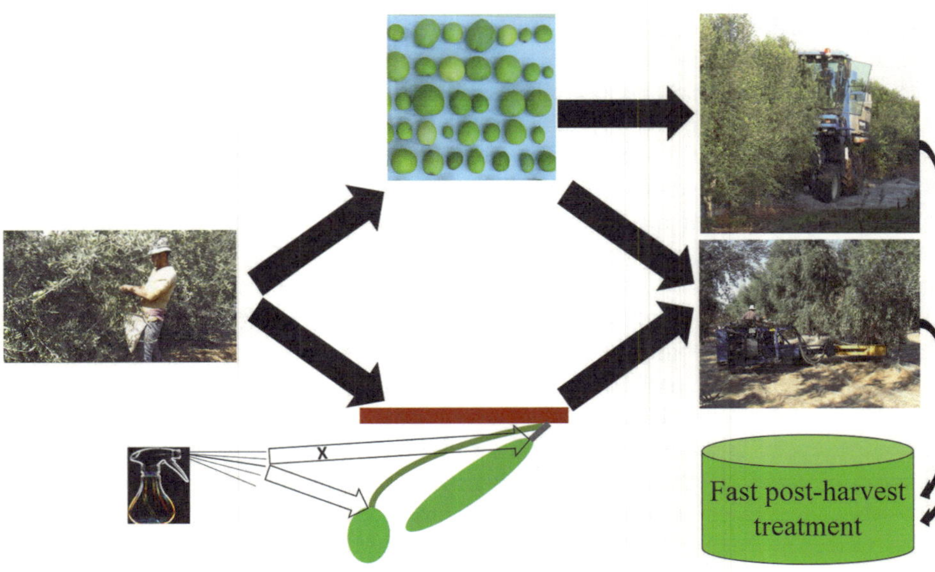

Figure 1.
In order to switch from hand to mechanical harvest of table olives several studies must be completed. The two main methods of mechanical harvesting of olives are use of a harvester or use of a trunk shaker. In order to use a harvester, we need to screen and identify (or develop) olive cultivars resistant to bruising, which can be harvested mechanically, without damage to the fruit. For use of a trunk shaker, we also need to identify or develop table olive cultivars resistant to bruising. This must be accompanied by a pre-harvest treatment to decrease fruit detachment force, while avoiding leaf abscission. In both mechanical harvesting methods, fast post-harvest treatment is crucial to avoiding defects and producing quality table olives.

The transition from traditional to fully mechanical systems is challenging not only for the individual farmer; investments in research and infrastructure are also required. New olive varieties adapted for intensive cultivation and super high density harvesting must be developed, suitable equipment must be purchased, and manpower trained. However, profits should cover these investments within a relatively short time (**Figure 1**).

The second area in which technical innovations are required is post-harvest treatments designed to decrease bruising of the fruit during harvest. These treatments should keep in mind the need to reduce the concentration of the polyphenols which cause the bitterness in untreated olives to desired levels. Natural fermentation has been confirmed as the best method for maintaining the high content of polyphenols and triacylglycerols necessary for preserving the nutritional and sensory qualities characteristic of table olives. Reducing processing time is another important goal of research, which must be integrated into the efforts to reduce bitterness and bruising of the fruit [4, 8].

Acknowledgements

We thank Yehuda Ben-Ari for valuable assistance in writing and editing this paper.

Conflicts of interest

The authors declare no conflict of interest.

Author details

Iris Biton, Dvora Namdar, Yair Mani and Giora Ben Ari*
Institute of Plant Sciences, ARO, The Volcani Center, Rishon LeZion, Israel

*Address all correspondence to: giora@agri.gov.il

References

[1] Galili E, Langgut D, Terral JF, Barazani O, Dag A, Kolska Horwitz L, et al. Early production of table olives at a mid-7th millennium BP submerged site off the Carmel coast (Israel). Scientific Reports. 2021;**11**:2218. DOI: 10.1038/s41598-020-80772-6

[2] Lanza B, Di Serio MG, Iannucci E, Russi F, Marfisi P. Nutritional, textural and sensorial characterisation of Italian table olives (Olea europaea L. cv. 'Intosso d'Abruzzo'). International Journal of Food Science & Technology. 2010;**45**:67-74. DOI: 10.1111/j.1365-2621.2009.02104.x

[3] IOC. World Catalogue of Olive Varieties. Príncipe de vergara, Madrid, Spain: International Olive Oil Council; 2013

[4] Conte P, Fadda C, Del Caro A, Urgeghe PP, Piga A. Table olives: An overview on effects of processing on nutritional and sensory quality. Food. 2020;**9**:514

[5] Garrido Fernández A, Fernández Díez MJ, Adams MR. Control methods. In: Table Olives Production and Processing. London: Chapman & Hall; 1997. pp. 461-480

[6] Bianchi G, Murelli C, Vlahov G. Surface waxes from olive fruits. Phytochemistry. 1992;**31**:3503-3506. DOI: 10.1016/0031-9422(92)83716-C

[7] Rodríguez G, Lama A, Rodríguez R, Jiménez A, Guillén R, Fernández-Bolaños J. Olive stone an attractive source of bioactive and valuable compounds. Bioresource Technology. 2008;**99**:5261-5269. DOI: 10.1016/j.biortech.2007.11.027

[8] Perpetuini G, Prete R, Garcia-Gonzalez N, Khairul Alam M, Corsetti A. Table olives more than a fermented food. Food. 2020;**9**:178

[9] Sánchez A-H, Ruiz-Barba JL, López-López A, Montaño A. Chapter 1- table olives: Types and trade preparations. In: Preedy VR, Watson RR, editors. Olives and Olive Oil in Health and Disease Prevention (Second Edition). San Diego: Academic Press; 2021. pp. 5-14. DOI: 10.1016/B978-0-12-819528-4.00019-5

[10] Lavee S, Avidan B, Ben-Ari G. Trends in breeding new olive varieties in Israel for quality and economic management. Agricultural Sciences. 2014;**05**(08):9. DOI: 10.4236/as.2014.58073

[11] Lanza B. Abnormal fermentations in table-olive processing: Microbial origin and sensory evaluation. Frontiers in Microbiology. 2013;**4**. Article 91. DOI: 10.3389/fmicb.2013.00091

[12] Ferguson L, Garcia SC. Transformation of an ancient crop: Preparing California 'Manzanillo' table olives for mechanical harvesting. HortTechnology. 2014;**24**:274-280. DOI: 10.21273/horttech.24.3.274

[13] Zipori I, Dag A, Tugendhaft Y, Birger R. Mechanical harvesting of table olives: Harvest efficiency and fruit quality. Horticultural Science. 2014;**49**:55-58. DOI: 10.21273/hortsci.49.1.55

[14] Jimenez-Jimenez F, Castro-Garcia S, Blanco-Roldan GL, González-Sánchez EJ, Gil-Ribes JA. Isolation of table olive damage causes and bruise time evolution during fruit detachment with trunk shaker. Spanish Journal of Agricultural Research. 2013;**11**:65-71. DOI: 10.5424/sjar/2013111-3399

[15] Zare F, Najafi G, Hashjin TT, Kermani AM, Ghiasi P. A new pneumatic harvester for improvement and facilitation the harvesting of the olive fruits. Advances in Horticultural Science. 2021;**35**:43-51. DOI: 10.36253/ahsc-9609

[16] Castro-Garcia S, Castillo F, Jiménez-Jiménez F, Gil-Ribes J, Blanco-Roldan G. Suitability of Spanish 'Manzanilla' table olive orchards for trunk shaker harvesting. Biosystems Engineering. 2014;**129**:388-395. DOI: 10.1016/j.biosystemseng.2014.11.012

[17] Jiménez-Jiménez F, Blanco-Roldan GL, Castillo F, Castro-Garcia S, Sola-Guirado R, Gil-Ribes JA. Table olives mechanical harvesting with trunk shakers: Orchard adaption and machine improvements. Chemical Engineering Transactions. 2015;**44**:271-276. DOI: 10.3303/CET1544046

[18] Morales-Sillero A, Rallo P, Jiménez MR, Casanova L, Suárez MP. Suitability of two table olive cultivars ('Manzanilla de Sevilla' and 'Manzanilla Cacereña') for mechanical harvesting in super high-density hedgerows. HortScience. 2014;**49**:1028-1033. DOI: 10.21273/hortsci.49.8.1028

[19] Goldental-Cohen S, Biton I, Many Y, Ben-Sason S, Zemach H, Avidan B, et al. Green olive Browning differ between cultivars. Frontiers in Plant Science. 2019;**10**:1260. DOI: 10.3389/fpls.2019.01260

[20] Hussein Z, Fawole OA, Opara UL. Harvest and postharvest factors affecting bruise damage of fresh fruits. Horticultural Plant Journal. 2020;**6**:1-13. DOI: 10.1016/j.hpj.2019.07.006

[21] Biton I, Doron-Faigenboim A, Jamwal M, Mani Y, Eshed R, Rosen A, et al. Development of a large set of SNP markers for assessing phylogenetic relationships between the olive cultivars composing the Israeli olive germplasm collection. Molecular Breeding. 2015;**35**:107. DOI: 10.1007/s11032-015-0304-7

[22] Nissim Y, Shloberg M, Biton I, Many Y, Doron-Faigenboim A, Zemach H, et al. High temperature environment reduces olive oil yield and quality. PLoS One. 2020;**15**:e0231956. DOI: 10.1371/journal.pone.0231956

[23] Nissim Y, Shlosberg M, Biton I, Many Y, Doron-Faigenboim A, Hovav R, et al. A high temperature environment regulates the olive oil biosynthesis network. Plants. 2020;**9**:1135

[24] Jiménez R, Rallo P, Suárez MP, Morales-Sillero AM, Casanova L, Rapoport HF. Cultivar susceptibility and anatomical evaluation of table olive fruit bruising. Acta Horticulturae. **924**:419-424. DOI: 10.17660/ActaHortic.2011.924.53

[25] Jiménez MR, Rallo P, Rapoport HF, Suárez MP. Distribution and timing of cell damage associated with olive fruit bruising and its use in analyzing susceptibility. Postharvest Biology and Technology. 2016;**111**:117-125. DOI: 10.1016/j.postharvbio.2015.07.029

[26] Casanova L, Corell M, Suárez MP, Rallo P, Martín-Palomo MJ, Jiménez MR. Bruising susceptibility of Manzanilla de Sevilla table olive cultivar under regulated deficit irrigation. Agricultural Water Management. 2017;**189**:1-4

[27] Casanova L, Corell M, Suárez MP, Rallo P, Martín-Palomo MJ, Morales-Sillero A, et al. Bruising response in 'Manzanilla de Sevilla' olives to RDI strategies based on water potential. Agricultural Water Management. 2019;**222**:265-273. DOI: 10.1016/j.agwat.2019.06.007

[28] Goldental-Cohen S, Burstein C, Biton I, Ben Sasson S, Sadeh A, Many Y, et al. Ethephon induced oxidative stress in the olive leaf abscission zone enables development of a selective abscission compound. BMC Plant Biology. 2017;**17**:87-87. DOI: 10.1186/s12870-017-1035-1

[29] Zipori I, Fishman A, Zelas ZB-B, Subbotin Y, Dag A. Effect of postharvest treatments of mechanically harvested 'Manzanilla' table olives on product quality. Postharvest Biology and Technology. 2021;**174**:111462. DOI: 10.1016/j.postharvbio.2021.111462

[30] Bonatsou S, Tassou CC, Panagou EZ, Nychas G-JE. Table olive fermentation using starter cultures with multifunctional potential. Microorganism. 2017;**5**:30. DOI: 10.3390/microorganisms5020030

Chapter 7

Olive Oil: Extraction Technology, Chemical Composition, and Enrichment Using Natural Additives

El Hassan Sakar and Said Gharby

Abstract

Virgin Olive oil (VOO) is considered the primary source of added fat in the Mediterranean diet. Its consumption is linked to numerous health-promoting properties along with its high energetic value. These properties are the results of various chemical compounds (fatty acids, tocopherols, polyphenols, etc.). VOO provides monounsaturated fatty acids, which lower total cholesterol and low-density lipoprotein cholesterol levels. VOO is obtained by three mechanical extraction processes, which can be classified into two systems that can be followed to extract olive oil from olives: the so-called traditional or discontinuous method, and the modern or continuous one. After the extraction of olive oil, its oxidative stability and chemical composition are subjected to deterioration especially when stored under inappropriate conditions (light, O_2, temperature, etc.). To deal with the problem, VOO enrichment using natural additives became an important practice to enhance VOO oxidative stability and its chemical composition. In this chapter, various aspects related to VOO extraction processes, chemical composition, stability oxidative and enrichment via natural additives will be reviewed and discussed in light of published literature.

Keywords: natural additives, chemical composition, enrichment, extraction technologies, olive oil

1. Introduction

Olive oil is one of a great interest in the vegetable oils world market. It is produced from the fruit of olive (*Olea europaea* L.). Virgin olive oil (VOO) is obtained exclusively by mechanical cold extraction [1]. It is not subjected to any chemical treatments apart from washing, decantation, centrifugation, and filtration. These processes may be carried out without refining, which makes the obtained oils highly appreciated by consumers thanks to their rich nutritional value, several health benefits, and unique organoleptic properties. Olive oil organoleptic and nutritional characteristics arise from noble compounds it contains. VOO composition consists of an unsaponifiable fraction (1 to 2%) along with essential unsaturated fatty acids contained in glyceridic

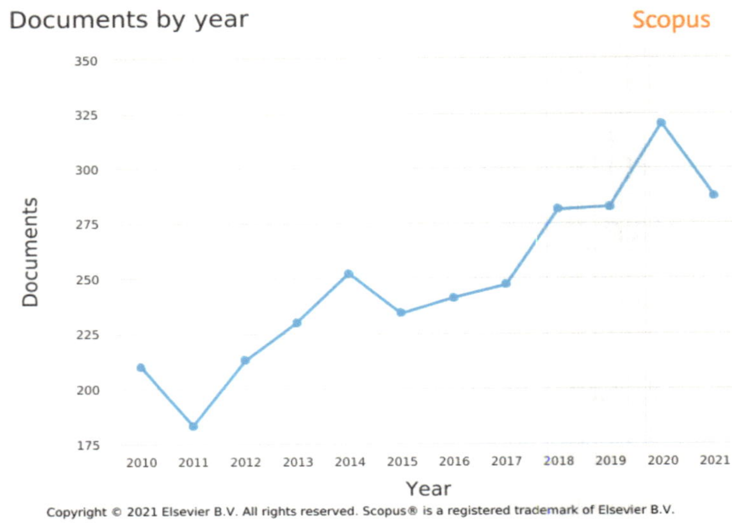

Figure 1.
Publication trends of olive oil composition (based on data retrieved from Scopus database).

fraction (98 to 99%) [2]. The composition of olive oil is well outlined in the literature. An updated analysis of the composition of olive oil reported in the literature is shown in **Figure 1**. In fact, on November 29th, 2021, the Scopus database was chosen to search for peer-reviewed literature regarding olive oil composition. The search string: ("olive oil*" AND "composition*") was utilized to extract bibliometric information from the Scopus online database. A total of 2980 publications were recovered through the literature search within the range of years from 2010 to 2021, 1892 of them representing about 65% are published in the interval of years from 2015 to 2021 (**Figure 1**).

2. Olive oil extraction technologies

Olive oil is made from fresh olives, which are extracted by mechanical processes [3]. Olive oil extraction technologies are summarized in **Figure 2**. There are two main olive oil extraction processes: traditional oil mills, and a relatively new extraction process known also by continuous mills and characterized by two or three phases [4, 5] **Figure 2**. All the above processes aim at separating the liquid oil phase from the other constituents of the fruit [6]. Likewise, olives should be processed as rapidly as possible after harvesting to reduce oxidation and preserve their quality [7].

Concerning the traditional press method, olive fruits liberated from leaves are washed, crushed using mill stones, and malaxed into a paste containing solid matter (core debris, epidermis, cell walls, etc.) and fluids (oil and vegetation water contained in the cells of olives). This is then spread on spherical mats [6]. Pressure with a hydraulic piston press is exerted then to obtain, firstly, a solid fraction (known as pomace) and, secondly, the mixture of oil and water is filled into a container and, eventually, the oil and water are then separated by gravity and collected through decantation [5]. The pressing process is the oldest method of obtaining olive oil [8]. Owing to lower production efficiency and high labor costs, during the last decade, the discontinuous pressing systems have widely been substituted by continuous systems, along with the development of centrifuge technology [4, 9]. After the steps of washing, crushing, and

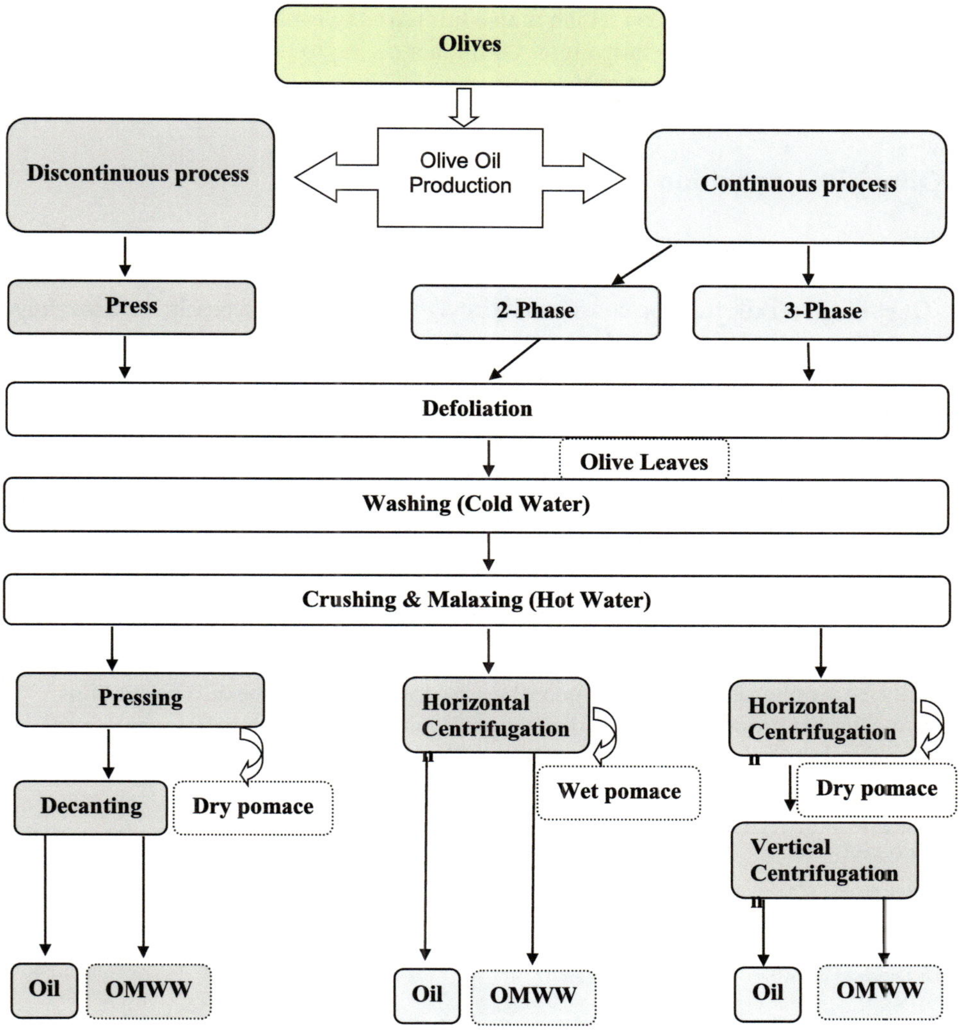

Figure 2.
Scheme of discontinuous and continuous extraction systems. OMWW = olive mill wastewater.

mixing, the mechanical extraction of the oil occurs mainly by a continuous process based on centrifugation using a decanter. The decanter centrifuge is equipped with a rotary bowl as well as a screw conveyor, which allows the processing of great quantities of olives in a short time [7]. Continuous separation systems can be divided into two-phase and three-phase systems, based on the decanter type used and the level of the phase of separation [9]. In the three-phase process, an additional amount of hot water is added to wash the oil, and then the three-phase decanter (insoluble solids, oil phase, and an aqueous phase), are separated following their density [7, 10]. Firstly, the solid wastes (insoluble solids), are separated from the remaining two phases in the decanter, and the liquid phases (oil phase as well as aqueous phase), are then subjected to vertical centrifugation to separate the olive oil from the olive mill wastewater [7].

Owing to the significant issue of wastewater produced, this three-phase system is preferred over the two-phase system since it is more eco-friendly [11]. This latest uses only a semi-liquid slurry (vegetation water along with insoluble solids) phase and the oil phase, a semi-liquid slurry, which is also known as two-phase olive mill waste [7]. This process has a reduced environmental impact owing to the reduced requirement of water as well as the amount of waste produced [7].

3. Olive oil composition

3.1 Bioactive compounds

Olive oil glyceridic fraction consists of triacylglycerols, diacylglycerols, monoacylglycerols and free fatty acids (FFA). Among them, 80% of them are unsaturated fatty acids. It is particularly rich in essential monounsaturated fatty acids (55–83% of oleic acid) and polyunsaturated fatty acids (2.5–21% of linoleic acid) [12]. The remaining fatty acids, apart from C16: 1, display an average value ranging from 0.3 to 3.5% (**Table 1** and **Figure 3**). Nevertheless, linolenic acid is a minority and its concentration is lower than 1% [12]. A low level of linolenic acid can be used to detect adulteration via some vegetable oils such as rapeseed and soybean oils [13]. Small quantities of saturated fatty acids also compose the triglycerides of olive oil: stearic acid (about 0.5–5%) and palmitic acid (about 7.5–20%). The remaining fatty acids (C17: 0, C17: 1, C20: 0, C20: 1, and C22: 0) are found to be of lower magnitudes. Since their concentrations are below 0.5% (**Table 1**). The unsaponified matter (about 1–2%) contains sterols, triterpene alcohols, tocopherol (mainly α-tocopherol), tocotrienol polyphenols, and squalene. The oil also contains a non-negligible proportion of volatile compounds. The total phytosterols content of VOO ranges between 100 and 200 mg/100 g. Also, 100 mg/100 g represents the inferior

Fatty acid [g/100 g]	Norm [12]	Physicochemical parameters	
Myristic acid [C14: 0]	≤ 0.2	Density [20°C]	0.906–0.919
Palmitic acid [C16: 0]	11.5–15	Refraction index [20°C]	1.463–1.472
Stearic acid [C18: 0]	4.3–7.2	Saponification value [mg of KOH/g]	184–196
Arachidic acid [C20: 0]	≤ 0.5	Iodine value [g (I2)/100 g]	75 to 94
Behenic acid [C22: 0]	≤ 0.2	Phytosterol [g/100 g]	(IOC 2021)
Σ *SFA*	15.8–23.1	Cholesterol	≤ 0.5
Palmitoleic acid [C16: 1]	≤ 0,2	Brassicasterol	≤ 0.1
Oleic acid [C18: 1]	43,0–49,1	Campesterol	≤ 4
Eicosenoic acid [C20: 1]	≤ 0.5	Stigmasterol	< Campesterol
Σ *MUFA*	43–49.8	Delta-7-stigmastenol	≤ 0.5
Linoleic acid [C18: 2]	29.3–36,0	Apparent beta-sitosterol	> 93
Linolenic acid [C18: 3]	≤ 0.3	Total sterol [mg/100 g]	≤ 220
Σ *PUFA*	29–36.3	Erythrodiol & Uvaol (% total sterols)	≤ 4.5

SFA-Saturated Fatty acids, MUFA-Monounsaturated fatty acids, PUFA-Polyunsaturated fatty acids: Apparent beta-sitosterol: beta-sitosterol +delta-5-avenasterol +delta-5-23-stigmastadienol +clerosterol + sitostanol +delta 5–24-stigmastadienol.

Table 1.
Physicochemical parameters, fatty acids, phytosterols, and tocopherols composition of olive oil.

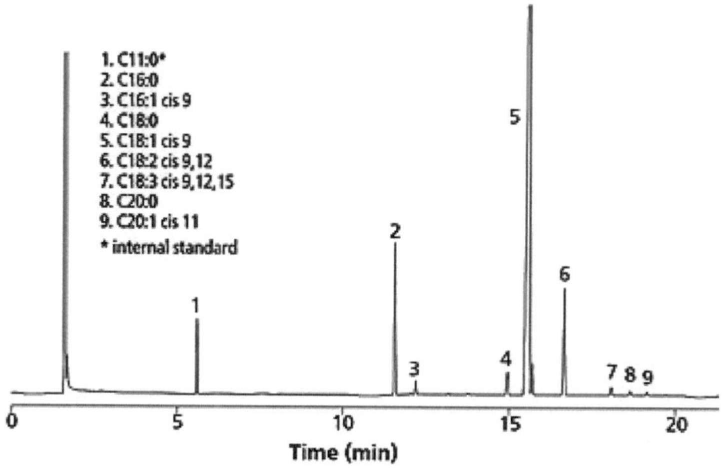

Figure 3.
Chromatogram of fatty acids.

limit set by the international olive council [12]. Apparent beta-sitosterol, (beta-sitosterol + delta-5-avenasterol + delta-5-23-stigmastadienol + clerosterol + sitostanol + delta 5–24-stigmastadienol) are the main compounds in the sterol fraction with a value more than 93% while β-sitosterol has the greatest relative percentage [14, 15] (**Figure 4**). VOO content also includes up to 4.5 g/100 g of total phytosterols [12]. The erythrodiol (5α-olean-12-ene-3β, 28-diol, homo-olestranol) in free and esterified forms and are the major triterpene di-alcohols found in olive oil [14], and their percentage reached up to 4.5% of the total content of sterols [12]. Moreover, four isoforms of tocopherols (α, β, γ, and δ-tocopherol) (**Figure 5**) and four tocotrienols (α, β, γ and δ- Tocotrienol) are present in olive oil. α-tocopherol is the main tocopherol found in olive oil, constituting more than 90% of the total tocopherol fraction [14]. Cunha et al. [16] reported that the proportions of tocopherols and tocotrienols ranged from 100 to 270 mg/kg in Portuguese olive oils [16]. Gharby et al. [47] found that the values of tocopherols varied from 150 to 250 mg/kg in three varieties ('Arbequina', 'Moroccan Picholine', and 'Picual') of olive oil [17]. Moreover, another study, based on the comparison of the tocopherol contents of olive oils from 4 different varieties harvested at different ripening periods found

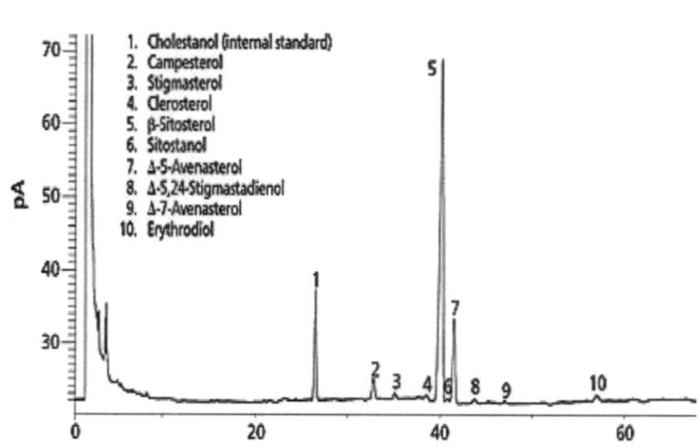

Figure 4.
Chromatogram of sterols.

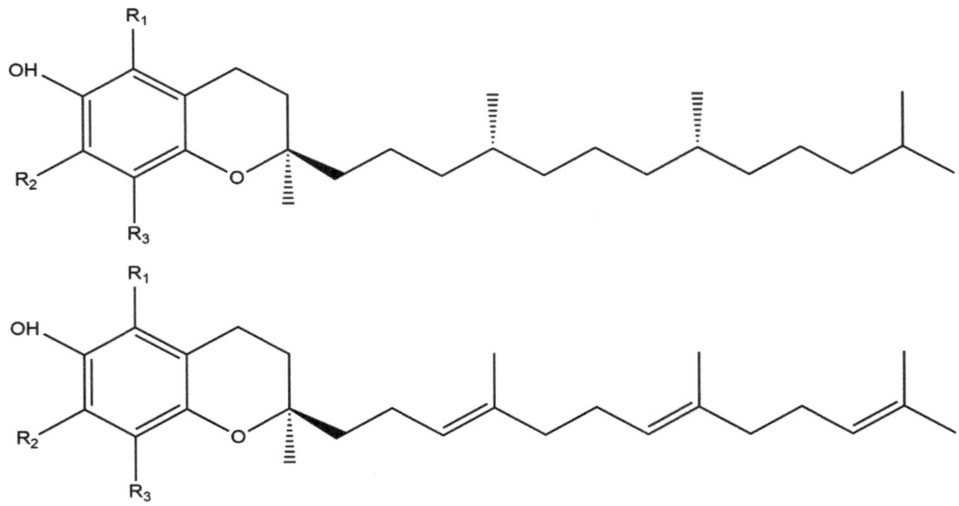

	R_1	R_2	R_3
alpha	CH_3	CH_3	CH_3
beta	CH_3	H	CH_3
gamma	H	CH_3	CH_3
delta	H	H	CH_3

Figure 5.
Tocopherols chemical structure.

that the α-tocopherol (major tocopherol) in oils obtained from olives composed of 130.54–180.43 mg/kg [18]. In general, tocopherol and tocotrienol levels in oil fluctuate with several factors such as harvest year, climatic conditions, storage time, extraction method, soil properties and spacing between olive trees [19]. Tocopherols possess a strong antioxidant power [20]. Together with tocopherols and tocotrienols, olive oil contains other antioxidant molecules such as polyphenolic compounds.

Many research works have demonstrated that the content of tocopherols in VOO is lower than that of argan oil [21–23].

The phenolic compounds are endowed to have a large scale of biological functions including stability to auto-oxidation, beneficial effects on human health [24]. About their well-known activities, olive oil polyphenols have been proven to possess an effective role in maintaining the organoleptic properties and the stability of olive oils [25].

Such bioactive compounds are extensively studied for their anti-inflammatory, antioxidant, neuroprotective, cardioprotective, antidiabetic, antimicrobial, and anticancer properties [26–29].

Franco et al. reported that phenolic compounds have a considerable increase during olive fruit growth. However, they are reduced when the fruits reach the maturation stage [30]. Khalatbary documented that the total phenolic content (TPC) in olive oils varies from 190 to 500 mg/kg [31]. In addition, in extra virgin olive oil, TPC commonly varies from 250 to 925 mg/kg [32]. Other factors including climatic conditions, variety, storage time, extraction conditions, soil properties, and analysis of polyphenolic compounds can lead to important variations in TPC [33]. Likewise, several classes of polyphenols are found in olive oils. These are presented as a separate class, to better understand the antioxidant phenolic chemistry of olive oil [33]. Finicelli et al. classify olive oil polyphenols following their chemical structure as follows [34]:

- Phenolic alcohols with a hydroxyl group are linked to an aromatic hydrocarbon group. The main constituents of this class are oleocanthal, hydroxytyrosol, and tyrosol (**Figure 6**) [26].

- Secoiridoids are phenolic compounds present in high amounts in olive oil in comparison to other plant species. The bitterness of extra virgin olive oil is a result of the content of secoiridoids [35].

- Lignans are chemically characterized by the aggregation of aromatic aldehydes. The pulp of the olives as well as the woody part of the seed contains lignans. These molecules are liberated into the oil during the process of extraction without biochemical changes [36].

- Flavonoids are chemically structured with two benzene rings attached via three linear carbon chains. The first flavonoids identified in VOO were flavones; their free forms, apigenin, and luteolin. They are the more abundant compounds [36].

- Hydroxyisocromans are the only two molecules characterized in commercial VOO. These compounds are produced via the HydroxyTyrosol reaction with benzaldehyde and vanillin [37].

- Phenolic acids are divided into two main classes: hydroxycinnamic acid along hydroxybenzoic acid [26].

The volatile fraction of VOOs has been reported to have about 280 different compounds [38]. The majority of volatile compounds are quickly developed during olive milling as a result of the disturbance of olive cells [39]. Although, Nardella et al. reported that most of the volatile compounds typical of olive oils are generated during

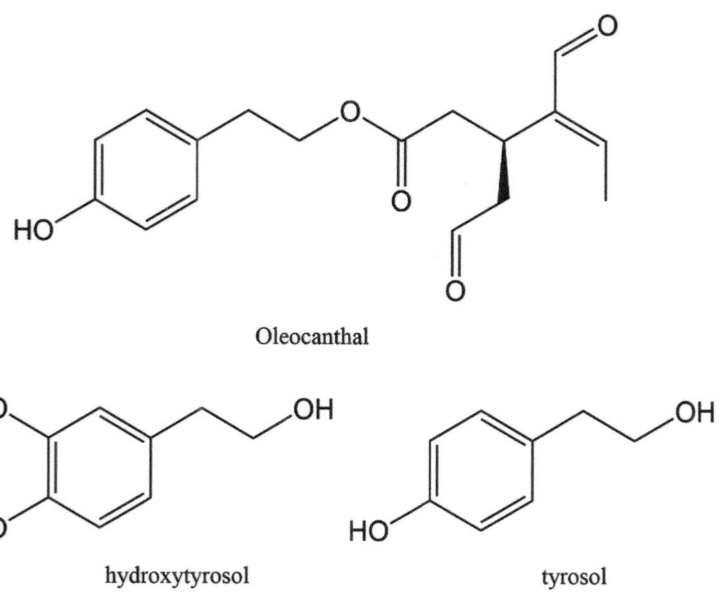

Figure 6.
Some phenolic alcohols present in olive oil.

malaxation due to the activation of particular pathways, in which the lipoxygenase (LOX) enzyme plays an essential role in producing a large quantity of C6 aldehydes, esters, and alcohols. These constitute almost all of the positive sensory marks in olive oils [40]. Such changes are initiated when olive tissues are affected, thereby enhancing the liberation of endogenous enzymes like hydroperoxide lyase and lipoxygenase [40].

Besides, several analytical techniques have been used to determine volatiles composition in olive oil. The main important are: GC (gas chromatography), HPLC (high-performance liquid chromatography), HPLC/MS (high-performance liquid chromatography/mass spectroscopy), IRMS (isotope ratio mass spectroscopy), ICP (inductively coupled plasma spectroscopy), NMR (nuclear magnetic resonance), SPME-GC/MS (solid-phase microextraction followed by gas chromatography/mass spectrometry, SNIF/NMR (specific natural isotopic fractionation nuclear magnetic resonance), SCIRA (stable carbon isotope ratio analysis), PTR/MS (proton transfer mass spectrometry) [41]. Fregapane et al. reported that the composition of volatiles may be affected significantly according to many factors such as cropping season, olive variety, harvest time, technological parameters, and agronomic conditions among other factors [42]. Ghanbari et al. reported that several chemical factors such as hydrophobicity, volatility, position, and functional groups type are reported to be directly linked to the odor degree of a given volatile component more than its content [38]. Theodosi et al. investigated correlations between the composition of volatiles of olive oil and altitude variation. The findings demonstrate that the total volatile compounds of 'Koroneiki' olive oil samples and altitude levels are negatively associated. The most important volatile compounds are alcohols, aldehydes, esters, and hydrocarbons [43].

3.2 Physicochemical parameters

Parameters routinely used to evaluate physicochemical properties of olive oil include density, iodine value, refractive index, saponification value along with unsaponifiable matter. For vegetable oils including olive oil, both density and refraction index depend on the temperature [44]. **Table 1** shows the ranges of the main physicochemical parameters of olive oil. The refractive index at 20°C varies in the range 1.463–1.472. At this same temperature, its density relative to water is between 0.906 and 0.919 [12]. The iodine value is a measure of the total number of double bonds found in an oil sample [1]. Olive oil displays an iodine value between 75 and 94 mg/100 g [17] (**Table 1**). This value is lower than that of argan oil (91–110 g I_2/100 g), and cactus seed oil (131.5 ± 0.5 g (I_2)/100 g) but higher than that of coconut oil (6.3–10.6 g (I_2)/100 g) [13]. High iodine value is associated with the greater number of double bonds and reduced oxidative stability [45]. The saponification value is a measure of the average chain lengths of fatty acids. An oil sample with shorter fatty acids has a high saponification value. Moreover, according to CODEX STAN 33, the saponification value of olive oil varies between 184 and 196 mg (KOH)/g.

4. Quality control of olive oil

Olive oil is subject to enormous analytical and sensory controls to assess its overall quality. These analyses evaluate the freshness of the oil regarding hydrolytic and oxidative alterations to ensure the conformity of products to their labels. For example, extra VOO by simple routine analyses (free fatty acids, peroxide value, specific extinction (E270 along with E232) and/or purity blending with other oils and contaminants. These criteria require detailed analyses (triglycerides contents, fatty acids, sterols, tocopherols, etc. ...). Organoleptic characteristics (taste, odor, color, etc. ...) also have to be taken into account.

As for other vegetable oils, the olive oil oxidation leads to natural phenomena alteration [46, 47]. This can be controlled since fruit harvest until oil storage. Because of oxidation, physicochemical parameters such as acidity, peroxide value and extinction specific at wavelength 270 (λ270 or λ270) have been selected as the backbone of olive oil quality determination by the International Olive Council [12]. Also, acidity of olive oil is classified into four grades: extra-virgin (Acidity < 0.8 g/100 g), fine-virgin (0.8 < Acidity < 2 g/100 g), ordinary virgin (2 < acidity < 3.3 g/100 g), and lampante olive oil (Acidity > 3.3) (**Table 2**) [12].

The variability of the extra VOO, acid value according to various parameters has been studied [47]. Oil oxidative state is examined from peroxide value and specific extinction coefficients (K232 or K270). These indicate the presence of primary and secondary oxidation products [1, 48]. The peroxide value of extra VOO oil must be below 20 mEq O_2/kg and specific extinction K232 < 2.5. The other two main indices used to evaluate the secondary oxidation products are the following: p-anisidine value and specific extinction K270 [1, 49]. The International Olive Council (IOC) has set 0.22 and 0.25 as a limit value for both the extra VOO and VOO, respectively [12].

Furthermore, along with oxidation and acidity concerns, the quantification of major compounds such as fatty acids (**Figure 3**), and minor compounds, like sterols (**Figure 4**), polyphenols, tocopherols, minerals elements, and other bioactive molecules, are also of great importance for the purity and for detection of olive oil adulteration, which is a complex problem. Owing to its high cost and demand, fraudsters blend VOO with cheaper edible oils (most often with sunflower and soybean oils) and sometimes with low-quality olive oil. Today, the problem exceeds the borders of the main producer countries and it tackles the international level market. In addition to known risks of commercializing a mixture of vegetable oils. There is another type of adulteration resulting from the mixing of relatively low and high-quality olive oil, and the outcome is a product, which is sold as “high quality extra VOO”. The control of adulteration, and authentication is of a crucial importance for the olive oil quality control. Codex Alimentarius (fats and oils), International Olive Council, and European Union Commission are dealing with the monitoring along with the regulation of VOO [50]. These international organizations have described the official control methods and have specified olive oil quality limits. Generally, all analytical techniques (chromatography, spectrophotometry, voltametric, differential scanning calorimetry), as well as several analytical methods, have been used to detect the adulteration of olive oil. Gas

Category olive oil	Acidity (g/100 g)	Peroxide index mEq O_2/Kg	Extinction specific at K_{232}	Extinction specific at K_{270}
1. Extra virgin olive oil	≤ 0.8	≤ 20	≤ 2.50	≤ 0.22
2. Fine Virgin olive oil	≤ 2.0	≤ 20	≤ 2.60	≤ 0.25
3. Ordinary Virgin olive oil	≤ 3.3	≤ 20	no limit	≤ 0.30
4. Lampante olive oil	> 3.3	no limit	no limit	no limit

Table 2.
Limits established for acidity, peroxide index and extinction specific (K_{232} and K_{270}) for each olive oil category.

chromatography (GC), which analyzes oil fatty acids profile, can be used to detect virgin oil purity by distinguishing it from other vegetable oils such as sunflower, soybean, walnut, rapeseed, and canola oils [51]. Moreover, HPLC-technique can be used, to calculate, the difference between the theoretical and experimental equivalent carbon number ($\Delta ECN42_{th}$). Likewise, the determination of phytosterols composition (namely campesterol Δ7-stigmasterol) using gas chromatography can be used to detect olive oil adulteration with low levels of cotton, corn, sunflower, soybean, and rapeseed oils [51]. In addition, Vietina et al. reported that the polymerase chain reaction (PCR) technique was demonstrated to be an efficient technique to detect VOO adulteration with cheaper vegetable oils by comparing their DNA melting profiles [52]. MS has also been used to detect the fraudulent presence of vegetable oils. Also, a lot of different techniques involving MS have been significantly developed, such as LC–MS, GC–MS, and MALDI-TOF/MS, which are of highly accurate identification [51]. Indeed, many other studies have also outlined the application of fluorescence spectroscopy, UV–Vis spectroscopy, [50] Fourier transform infrared spectroscopy [53] mid-infrared (MID) or near-infrared spectroscopy (NIR) [54] and Raman spectroscopy [55] for authentication and detection of adulteration of vegetables oil present in VOO [50]. Otherwise, differential scanning calorimetry (DSC) has also been used to detect argan oil purity by discriminating it from sunflower, high oleic sunflower as well as refined hazelnut oil [50]. Apetrei and Apetrei have investigated the use of the voltametric method based on modified EO

Negative attributes		
Fusty/muddy sediment	The characteristic flavor of oil obtained from olives stocked in such a way that they have developed an enhanced stage of anaerobic fermentation,	
Musty-humid-earthy	Flavor characteristic of oils from fruits in which a lot of fungi and yeasts have been developed or from olives picked up with mud or earth and not been previously cleaned.	
Winey- vinegary	The flavor characteristic of some oils reminiscent of vinegar or wine.	
Acid-sour	The flavor is primarily caused by the formation of ethyl acetate, ethanol, and acetic acid.	
Rancid	Flavor of oils that have been submitted to an intense oxidation process	
Frostbitten olives (wet wood)	Flavor characteristic of oils obtained from olives that have been damaged by frost on the tree.	
Positive attributes		
Fruity	Characteristic of the oil that varies according to the variety and is obtained from fresh olives, ripe or not.	
Bitter	Primary taste characteristic of oil extracted from green olives or olives that are becoming colored.	
Pungent	Characteristic of oils obtained at the beginning of the crop year, mostly from olives that are not ripe yet.	
	Median of defect (Md)	**Fruity median (Mf)**
1. Extra virgin oil	Md = 0	Mf > 0
2. Virgin olive oil	Md ≤ 3.5	Mf > 0
3. Lampante oil	Md > 3.5	—

Table 3.
Organoleptic attributes of olive oil.

carbon paste-based sensors to determine the adulteration of VOO with soybean and sunflower oils [56].

On the other hand, identification of contaminants is one of the multiple checks that must be performed on oils. Vegetable oils have limited values for aromatic hydrocarbons polycyclics (PAHs), heavy metals, mycotoxins, phthalates, and pesticides. Although, the physicochemical characterization of olive oil is an essential step, it is not sufficient and organoleptic characteristics along with the above-mentioned supplementary analyses are required for a full picture of olive oil quality [1].

To satisfy consumers, organoleptic characteristics (color, taste, smell, etc.) must be taken into account. This is particularly important for olive oil. The organoleptic analysis is an essential step for successful food marketing. It is an integral part of evaluating olive oil. IOC has established a procedure to evaluate the organoleptic characteristics of VOO according to COI/T.20/Doc. [12] No 15/Rev. 102,018. It has classified such characteristics into positive and negative attributes as highlighted in **Tables 2** and **3**.

5. Olive oil enrichment with natural additives

Oxidation of lipids including oils is a major concern to food industries [57, 58]. While, vegetable oils are endowed with a wide variety of endogenous antioxidants (pigments, vitamins, tocols, phenols, etc.), the use of exogenous antioxidants is widely practiced to enhance oxidative stability. In this regard, synthetic antioxidants such as butylated hydroxyanisole (BHA), butylated hydroxytoluene (BHT), tertiary butylhydroquinone (TBHQ), as well propyl gallate are commercially used to extend oils' shelf life by delaying or even hindering lipids degradation. These molecules are considered as Generally Recognized as Safe (GRAS) preservatives with a concentration limit of 0.02% in oils and fats [59]. In contrast, some reports associated these molecules with health risks because of carcinogenesis, leading to a restriction of the use of the GRAS list and a reduction of their utilization in different countries [59]. For this reason, natural antioxidants are a good alternative to replace the synthetic ones in preserving vegetable oils including olive oil [59–61]. An overview of factors involved in the balance of antioxidants and pro-oxidants as well as synthetic and natural antioxidants are summarized in **Figure 7**.

Natural extracts sourced from various plant parts (peel, fruit, leaf, flower, and root) from different aromatic and medicinal herbs, agri-food residues and by-products were investigated for their antioxidant power as well as their use for the enrichment of olive oil with an emphasis on improving oxidative stability. Such natural extracts were proved to have a wide range of bioactive compounds were identified. These are mainly carotenoids and phenols [62, 63]. Promising results were obtained regarding the improvement of oxidative stability and shelf life of olive oil. Regarding the antioxidant activity of synthetic and natural additives, several mechanisms are involved. They act as free radical scavengers, inactivators of peroxides as well as other reactive oxygen species (ROS), singlet oxygen quenchers, metal ion chelators, quenchers of secondary oxidation products, and inhibitors of pro-oxidative enzymes, among other compounds [64]. Following these authors, antioxidants can be classified, based on their mode of action, into primary antioxidants. These break the oxidation chain reaction through scavenging free radical intermediates, however secondary antioxidants delay or even prevent oxidation through suppression of oxidation initiator, accelerators or regeneration of primary antioxidants.

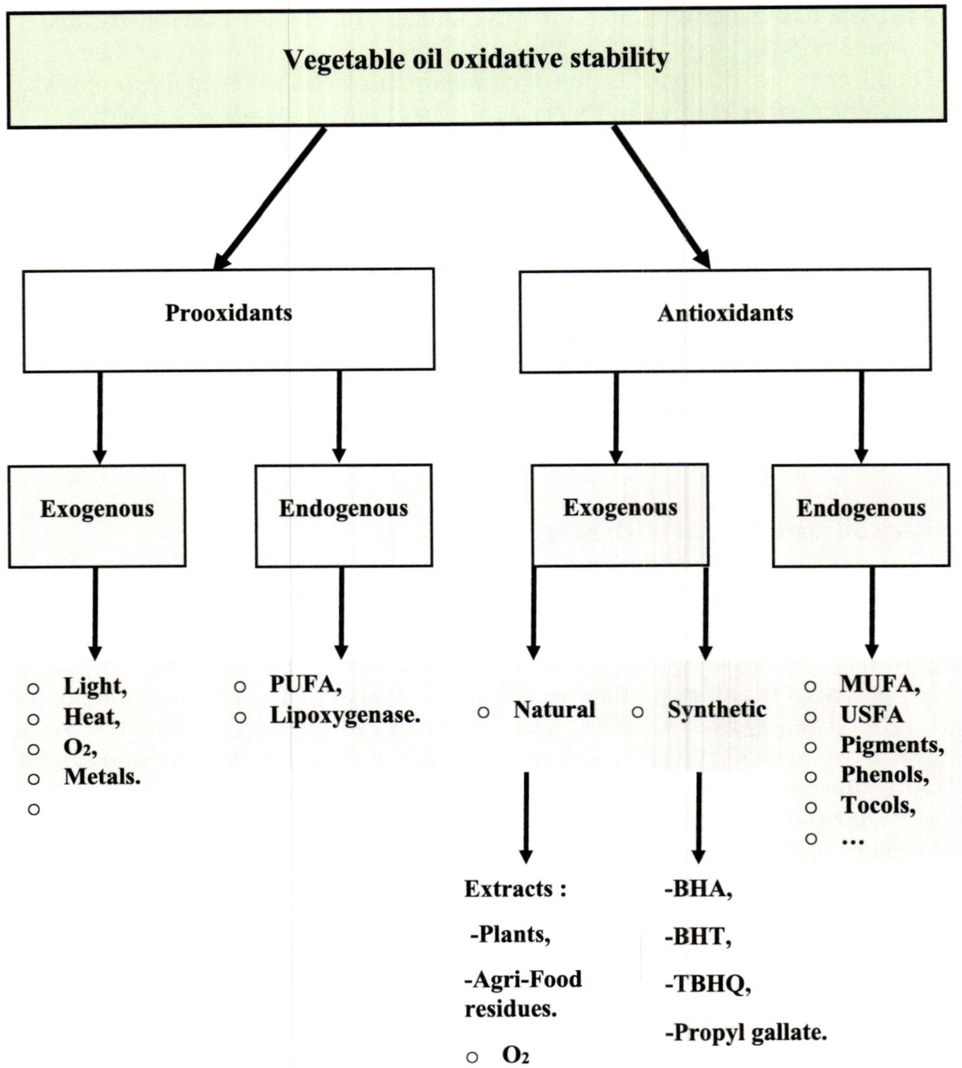

Figure 7.
An overview of factors involved in olive oil oxidative stability as well as natural and synthetic antioxidants. BHA, butylated hydroxyanisole; BHT, butylated hydrolxytoluene; TBHQ, tertiary butylhydroquinone; MUFA, monounsaturated fatty acids, and PUFA, polyunsaturated fatty acids.

6. Conclusions

Olive oil is an important food in the Mediterranean diet. Its importance and nutritional value arise from chemical composition. Its richness in essential fatty acids is behind their health-promoting properties. A set of other minor compounds such as polyphenols and tocopherols act as antioxidants which are directly associated with oxidative stability and shelf life of olive oil on one hand as well as human health on the other hand. Along with these endogenous antioxidants, olive oil quality can be enhanced through natural antioxidants extracted from herbs and agri-food residues.

Acknowledgements

This work has no acknowledgments.

Conflict of interest

The authors declare no conflict of interest.

Author details

El Hassan Sakar[1*] and Said Gharby[2]

1 Laboratory of Applied Botany, FS, Abdelmalek Essaadi University, Tetuan, Morocco

2 Laboratory Biotechnology, Materials and Environment (LBME), Polydisciplinary Faculty of Taroudant, Ibn Zohr University, Morocco

*Address all correspondence to: e.sakar@uae.ac.ma

References

[1] Gharby S, Hajib A, Ibourki M, et al. Induced changes in olive oil subjected to various chemical refining steps: A comparative study of quality indices, fatty acids, bioactive minor components, and oxidation stability kinetic parameters. Chemical Data Collections. 2021;**33**:100702. DOI: 10.1016/j.cdc.2021.100702

[2] Flori L, Donnini S, Calderone V, et al. The nutraceutical value of olive oil and its bioactive constituents on the cardiovascular system. Focusing on Main strategies to slow down its quality decay during production and storage. Nutrients. 2019;**11**:1962. DOI: 10.3390/nu11091962

[3] Beltrán G, Bejaoui MA, Sánchez-Ortiz A, Jiménez A. Water addition during oil extraction affects on virgin olive oil ethanol content, quality and composition. European Journal of Lipid Science and Technology. 2021;**123**:1-7. DOI: 10.1002/ejlt.202000400

[4] Domingues E, Fernandes E, Gomes J, et al. Olive oil extraction industry wastewater treatment by coagulation and Fenton's process. Journal of Water Process Engineering. 2021;**39**:101818. DOI: 10.1016/j.jwpe.2020.101818

[5] Gullón P, Gullón B, Astray G, et al. Valorization of by-products from olive oil industry and added-value applications for innovative functional foods. Food Research International. 2020;**137**:109683. DOI: 10.1016/j.foodres.2020.109683

[6] Goldsmith CD, Stathopoulos CE, Golding JB, Roach PD. Fate of the paorlic compounds during olive oil production with the traditional press method. International Food Research Journal. 2014;**21**:101-109

[7] Souilem S, El-Abbassi A, Kiai H, et al. Olive oil production sector: environmental effects and sustainability challenges in Olive mill waste. In: Galanakis CM, editor. London: Academic Press; 2017. p. 1-28. DOI: 10.1016/B978-0-12-805314-0.00001-7

[8] Khdair AI, Ayoub S, Abu-Rumman G. Effect of pressing techniques on olive oil quality. American Journal of Food Technology. 2015;**10**:176-183. DOI: 10.3923/ajft.2015.176 183

[9] Hocaoglu SM, Gursoy Haksevenler BH, Basturk I, et al. Assessment of technology modification for olive oil sector through mass balance: A case study for Turkey. Journal of Cleaner Production. 2018;**188**:786-795. DOI: 10.1016/j.jclepro.2018.04.020

[10] Cheremisinoff PN. Centrifugal Separation. In: Cheremisinoff PN, editor. Handbook of Water and Wastewater Treatment Technology. 1st ed. Boca Raton: Routledge; 1995. pp. 339-372. DOI: 10.1201/9780203752494

[11] Nunes MA, Pimentel FB, Costa ASG, et al. Olive by-products for functional and food applications: Challenging opportunities to face environmental constraints. Innovative Food Science and Emerging Technologies. 2016; **35**:139-148. DOI: 10.1016/j.ifset.2016.04.016

[12] IOC/T.15/NC No 3/Rév. NORME COMMERCIALE APPLICABLE AUX HUILES D'OLIVE ET AUX HUILES DE GRIGNONS D'OLIVE. 2019; 1-17

[13] Gharby S, Ravi HK, Guillaume D, et al. 2-methyloxolane as alternative solvent for lipid extraction and its effect on the cactus (Opuntia ficus-indica L.) seed oil fractions. OCL. 2020;**27**:27. DOI: 10.1051/ocl/2020021

[14] Jimenez-Lopez C, Carpena M, Lourenço-Lopes C, et al. Bioactive compounds and quality of extra virgin olive oil. Food. 2020;**9**:1014

[15] Gharby S, Harhar H, Farssi M, Abdelaziz Ait Taleb DG, Laknifli A. Influence of roasting olive fruit on the chemical composition and polycyclic aromatic hydrocarbon content of olive oil. OCL - Oilseeds and fats, Crops and Lipids. 2018;**25**:A303. DOI: 10.1051/ocl/2018013

[16] Cunha SC, Amaral JS, Fernandes JO, MBPP O. Quantification of tocopherols and tocotrienols in Portuguese olive oils using HPLC with three different detection systems. Journal of Agricultural and Food Chemistry. 2006;**54**:3351-3356. DOI: 10.1021/jf053102n

[17] Gharby S, Harhar H, El Monfalouti H, et al. Chemical and oxidative properties of olive and argan oils sold on the Moroccan market. A comparative study. Mediterranean. Journal of Nutrition and Metabolism. 2012;**5**:31-38. DOI: 10.1007/s12349-011-0076-5

[18] Dağdelen A, Tümen G, Özcan MM, Dündar E. Determination of tocopherol contents of some olive varieties harvested at different ripening periods. Natural Product Research. 2012;**26**:1454-1457. DOI: 10.1080/14786419.2011.605364

[19] Rodrigues N, Casal S, Peres AM, et al. Effect of olive trees density on the quality and composition of olive oil from cv. Arbequina. Scientia Horticulturae. 2018;**238**:222-233. DOI: 10.1016/j.scienta.2018.04.059

[20] Charrouf Z, Guillaume D. The argan oil project: Going from utopia to reality in 20 years. OCL - Oilseeds and fats, Crops and Lipids. 2018;**25**:1-5. DOI: 10.1051/ocl/2018006

[21] Zarrouk A, Martine L, Grégoire S, et al. Profile of fatty acids, tocopherols, Phytosterols and polyphenols in Mediterranean oils (Argan oils, olive oils, Milk thistle seed oils and Nigella seed oil) and evaluation of their antioxidant and Cytoprotective activities. Current Pharmaceutical Design. 2019;**25**:1791-1805. DOI: 10.2174/13816128256661907 05192902

[22] El Kharrassi Y, Maata N, Mazri MA, et al. Chemical and phytochemical characterizations of argan oil (Argania spinosa L. skeels), olive oil (Olea europaea L. cv. Moroccan picholine), cactus pear (Opuntia megacantha salm-dyck) seed oil and cactus cladode essential oil. Journal of Food Measurement and Characterization. 2018;**12**:747-754. DOI: 10.1007/s11694-017-9688-x

[23] Marfil R, Giménez R, Martínez O, et al. Determination of polyphenols, tocopherols, and antioxidant capacity in virgin argan oil (Argania spinosa, Skeels). European Journal of Lipid Science and Technology. 2011;**113**:886-893. DOI: 10.1002/ejlt.201000503

[24] Robles-Almazan M, Pulido-Moran M, Moreno-Fernandez J, et al. Hydroxytyrosol: Bioavailability, toxicity, and clinical applications. Food Research International. 2018;**105**:654-667. DOI: 10.1016/j.foodres.2017.11.053

[25] Bayram B, Esatbeyoglu T, Schulze N, et al. Comprehensive analysis of polyphenols in 55 extra virgin olive oils by HPLC-ECD and their correlation with antioxidant activities. Plant Foods for Human Nutrition. 2012;**67**:326-336. DOI: 10.1007/s11130-012-0315-z

[26] Rodríguez-López P, Lozano-Sanchez J, Borrás-Linares I, et al. Structure–biological activity relationships of extra-virgin olive oil phenolic compounds: Health properties and bioavailability. Antioxidants. 2020;**9**:1-17. DOI: 10.3390/antiox9080685

[27] Marković AK, Torić J, Barbarić M, Brala CJ. Hydroxytyrosol, tyrosol and derivatives and their potential effects on human health. Molecules. 2019;**24**:2001. DOI: 10.3390/molecules24102001

[28] Rigacci S, Stefani M. Nutraceutical properties of olive oil polyphenols. An itinerary from cultured cells through animal models to humans. International Journal of Molecular Sciences. 2016;**17**: 1-28. DOI: 10.3390/ijms17060843

[29] Cicerale S, Lucas LJ, RSJ K. Antimicrobial, antioxidant and anti-inflammatory phenolic activities in extra virgin olive oil. Current Opinion in Biotechnology. 2012;**23**:129-135. DOI: 10.1016/j.copbio.2011.09.006

[30] Franco MN, Galeano-Díaz T, López Ó, et al. Phenolic compounds and antioxidant capacity of virgin olive oil. Food Chemistry. 2014;**163**:289-298. DOI: 10.1016/j.foodchem.2014.04.091

[31] Khalatbary AR. Olive oil phenols and neuroprotection. Nutritional Neuroscience. 2013;**16**:243-249. DOI: 10.1179/1476830513Y.0000000052

[32] Kotsiou K, Tasioula-Margari M. Monitoring the phenolic compounds of Greek extra-virgin olive oils during storage this work is in memory of Maria Tasioula-Margari, who was the thesis supervisor. Food Chemistry. 2016;**200**: 255-262. DOI: 10.1016/j.foodchem.2015.12.090

[33] Zeb A. A comprehensive review on different classes of polyphenolic compounds present in edible oils. Food Research International. 2012;**143**:110312. DOI: 10.1016/j.foodres.2021.110312

[34] Finicelli M, Squillaro T, Galderisi U, Peluso G. Polyphenols, the healthy brand of olive oil. Insights and Perspectives. Nutrient. 2021;**13**(11):3831. DOI: 10.3390/nu13113831

[35] Mateos R, Cert A, Carmen Pérez-Camino M, García JM. Evaluation of virgin olive oil bitterness by quantification of Secoiridoid derivatives. JAOCS, Journal of the American Oil Chemists' Society. 2004;**81**:71-75. DOI: 10.1007/s11746-004-0859-x

[36] de Torres A, Espínola F, Moya M, et al. Assessment of phenolic compounds in virgin olive oil by response surface methodology with particular focus on flavonoids and lignans. LWT - Food Science and Technology. 2018;**90**:22-30. DOI: 10.1016/j.lwt.2017.12.003

[37] de Souza PAL, Marcadenti A, Portal VL. Effects of olive oil phenolic compounds on inflammation in the prevention and treatment of coronary artery disease. Nutrients. 2017;**9**:1087. DOI: 10.3390/nu9101087

[38] Ghanbari R, Anwar F, Alkharfy KM, et al. Valuable nutrients and functional bioactives in different parts of olive (Olea europaea L.) - a review. International Journal of Molecular Sciences. 2012;**13**(3):3291-3340. DOI: 10.3390/ijms13033291

[39] Campestre C, Angelini G, Gasbarri C, Angerosa F. The compounds responsible for the sensory profile in monovarietal virgin olive oils. Molecules. 2017;**22**:1-28. DOI: 10.3390/molecules22111833

[40] Nardella M, Moscetti R, Nallan Chakravartula SS, Bedini G, Massantini R. A review on high-power ultrasound-assisted extraction of olive oils: Effect on oil yield, quality, chemical composition and consumer perception. Food. 2021;**10**: 2743

[41] Kosma I, Badeka A, Vatavali K, et al. Differentiation of Greek extra virgin olive oils according to cultivar based on volatile compound analysis and fatty acid composition. European Journal of Lipid

Science and Technology. 2016;**118**:849-861. DOI: 10.1002/ejlt.201500293

[42] Fregapane G, Salvador MD. Chemical and sensory characteristics of extra virgin olive oils produced in Central Iberian Peninsula under the protected designation of origin quality scheme. European Journal of Lipid Science and Technology. 2019;**121**:1-23. DOI: 10.1002/ejlt.201800134

[43] Theodosi S, Kosma IS, Badeka AV. Quality characteristics of Koroneiki olive oil from Zakynthos island (Greece) and differentiation depending on the altitude level. European Food Research and Technology. 2021;**247**:1235-1248. DOI: 10.1007/s00217-021-03705-1

[44] Bijla L, Aissa R, Bouzid HA, et al. Spent coffee ground oil as a potential alternative for vegetable oil production: Evidence from oil content, lipid profiling, and physicochemical characterization. Biointerface Research in Applied Chemistry. 2022;**12**:6308-6320. DOI: 10.33263/BRIAC125.63086320

[45] Gharby S, Harhar H, Bouzoubaa Z, et al. Chemical characterization and oxidative stability of seeds and oil of sesame grown in Morocco. Journal of the Saudi Society of Agricultural Sciences. 2017;**16**:105-111. DOI: 10.1016/j.jssas.2015.03.004

[46] Choe E, Min DB. Mechanisms and factors for edible oil oxidation. Comprehensive Reviews in Food Science and Food Safety. 2006;**5**:169-186. DOI: 10.1111/j.1541-4337.2006.00009.x

[47] Gharby S, Harhar H, Guillaume D, et al. Oxidative stability of edible argan oil: A two-year study. LWT - Food Science and Technology. 2011;**44**:1-8. DOI: 10.1016/j.lwt.2010.07.003

[48] Harhar H, Gharby S, Eddine B, El Monfalouti H. Long argan fruit drying time is detrimental for argan oil quality. Natural Product Communications. 2010;**5**:1934578X1000501122. DOI: 10.1177/1934578X1000501122

[49] Harhar H, Gharby S, Guillaume D, Charrouf Z. Effect of argan kernel storage conditions on argan oil quality. European Journal of Lipid Science and Technology. 2010;**112**:915-920. DOI: 10.1002/ejlt.200900269

[50] Meenu M, Cai Q, Xu B. A critical review on analytical techniques to detect adulteration of extra virgin olive oil. Trends in Food Science and Technology. 2019;**91**:391-408. DOI: 10.1016/j.tifs.2019.07.045

[51] Salah WA, Nofal M. Review of some adulteration detection techniques of edible oils. Journal of the Science of Food and Agriculture. 2021;**101**:811-819. DOI: 10.1002/jsfa.10750

[52] Vietina M, Agrimonti C, Marmiroli N. Detection of plant oil DNA using high resolution melting (HRM) post PCR analysis: A tool for disclosure of olive oil adulteration. Food Chemistry. 2013;**141**:3820-3826. DOI: 10.1016/j.foodchem.2013.06.075

[53] Maggio RM, Cerretani L, Chiavaro E, et al. A novel chemometric strategy for the estimation of extra virgin olive oil adulteration with edible oils. Food Control. 2010;**21**:890-895. DOI: 10.1016/j.foodcont.2009.12.006

[54] Mendes TO, da Rocha RA, BLS P, et al. Quantification of extra-virgin olive oil adulteration with soybean oil: A comparative study of NIR, MIR, and Raman spectroscopy associated with Chemometric approaches. Food Analytical Methods. 2015;**8**:2339-2346. DOI: 10.1007/s12161-015-0121-y

[55] Farley C, Kassu A, Bose N, et al. Short distance standoff Raman detection of extra virgin olive oil adulterated with canola and grapeseed oils. Applied Spectroscopy. 2017;**71**:1340-1347. DOI: 10.1177/0003702816681796

[56] Apetrei IM, Apetrei C. Detection of virgin olive oil adulteration using a voltammetric e-tongue. Computers and Electronics in Agriculture. 2014;**108**:148-154. DOI: 10.1016/j.compag.2014.08.002

[57] Domínguez R, Pateiro M, Gagaoua M, et al. A comprehensive review on lipid oxidation in meat and meat products. Antioxidants. 2019;**8**:1-31. DOI: 10.3390/antiox8100429

[58] Ghorbani Gorji S, Smyth HE, Sharma M, Fitzgerald M. Lipid oxidation in mayonnaise and the role of natural antioxidants: A review. Trends in Food Science and Technology. 2016;**56**:88-102. DOI: 10.1016/j.tifs.2016.08.002

[59] Taghvaei M, Jafari SM. Application and stability of natural antioxidants in edible oils in order to substitute synthetic additives. Journal of Food Science and Technology. 2015;**52**:1272-1282. DOI: 10.1007/s13197-013-1080-1

[60] Ungureanu CRM, Poiana MA, Cocan I, et al. Strategies to improve the thermo-oxidative stability of sunflower oil by exploiting the antioxidant potential of blueberries processing byproducts. Molecules. 2020;**25**:1-21. DOI: 10.3390/molecules25235688

[61] Odeh D, Kraljić K, Skukan AB, Škevin D. Oxidative stability, microbial safety, and sensory properties of flaxseed (Linum usitatissimum l.) oil infused with spices and herbs. Antioxidants. 2021;**10**:785. DOI: 10.3390/antiox10050785

[62] Yang Y, Song X, Sui X, et al. Rosemary extract can be used as a synthetic antioxidant to improve vegetable oil oxidative stability. Industrial Crops and Products. 2016;**80**:141-147. DOI: 10.1016/j.indcrop.2015.11.044

[63] Blasi F, Cossignani L. An overview of natural extracts with antioxidant activity for the improvement of the oxidative stability and shelf life of edible oils. PRO. 2020;**8**:956. DOI: 10.3390/PR8080956

[64] Fereidoon S, Ying Z. Lipid oxidation and improving the oxidative stability. Chemical Society Reviews. 2010;**39**:4067-4079. DOI: 10.1039/b922183m

Chapter 8

Harmony (QHV): Practical Experiences with an Additional Sensory Criterion for the Quality Assessment of the Grade "Extra Virgin Olive Oil"

Dietrich Oberg

Abstract

Well over 100 laboratories and tester groups worldwide check about 2,4 Mio tons of olive oil according to the Reg. (EWG) 2568/91 in its actual version or according to the method COI/T20 Doc.No.15. Pursuant to the regulation around 50% is EVOO. In order to evaluate these versatile typical sensory characteristics, the additional criterion "Quantified Harmony Value" (QHV) was developed. The QHV assessment is evaluated for all EVOO with special focus on the mass market. The method is based on the relationship of smell to taste and the interplay with the attributes such as bitter and pungent, on the cleanliness of all attributes and flavor, on the evaluation of complexity and persistency. All these findings play an important role in olive oil competitions—organized exclusively for a small number of premium products. But the higher the quality, the lower the quantity. Therefore, large quantities of EVOO need as well objective differentiations. But for the largest share of EVOO produced for millions of consumers, there are officially no distinguishing criteria. The additional criteria QHV close this gap and make it objectively possible to discriminate all oils in the EVOO category at commercial level. The study at hand explains the relevant method and shows the positive development of the last 10 years using the example of the importing country Germany.

Keywords: grade extra virgin, harmony criteri0n, method, profiling, experiences, commercial benefits

1. Introduction

Regardless of the quantities harvested, the average percentage of virgin olive oil of the highest category (EVOO) in the three main producing countries, Spain, Italy, and Greece, is estimated at about 50% per year [EC Statistics, 1998]. The remainder has chemical or sensory defects that, depending on their intensity, must be classified as virgin olive oil (VOO), ordinary virgin olive oil (OVOO), or lampante virgin olive oil (LVOO) according to EEC-Regulation 2568/91 or International Olive Council (IOC) Standard. Olive oil is one of the most controlled vegetable oils in the world. Nevertheless, the EU classified olive oil as one of the most adulterated food products, although it is subject to a number of EC regulations [1].

Olive oil is one of the few fruit oils, along with avocado oil and palm oil. Most vegetable oils known to us are seed oils. The olive fruit goes through many important steps on its way "from tree to bottle" that can affect its quality. The fruit itself belongs to a range of well over 500 different varieties. They differ in the size of the fruit, size and type of leaves, ripening, kernel/pulp ratio, amount of oil, unusual minor components, and among other things, especially in taste. Some varieties are used only as table olives, others for both purposes, but most are processed into oil. In the plantations, the olive trees are planted traditionally, intensively, and super intensively. The spacing between trees has decreased from 7.50 × 7.50 m (traditional) to 3.50 × 3.50 m (intensive) to trellis-like cultivation as in the case of wine (super intensive). The main reasons for this type of cultivation are an improvement in yield and the use of harvesting machinery. Harvesting is the most expensive part. The most important factor for a good-quality oil is the right time of harvesting or the optimal ripeness of the fruit. After harvesting, the quality of the fruit can only decrease, as fermentative degradation processes begin immediately after harvesting. Picking the olives by hand is often the only way to harvest the fruit, as the cultivated areas make it impossible to use harvesting machines. It is important that the fruit is not damaged during harvesting, and the waiting time before pressing or extraction in the decanter should not exceed 24–48 hours. Pressing requirements range from high quantities of olives in a correspondingly short time to achieving a particular quality—regardless of quantity—with pressing temperatures or conditions during extraction in the decanter, high polyphenol content, and excellent aroma and flavor.

Ultimately, each of the above processing steps results in a wide range of different qualities in terms of aroma (smell and taste), complexity, nutritional composition, and oxidation stability.

When the oil is ready for marketing, each producer follows his own way of marketing. Depending on the size of the crop, some producers handle storage, bottling, labeling,

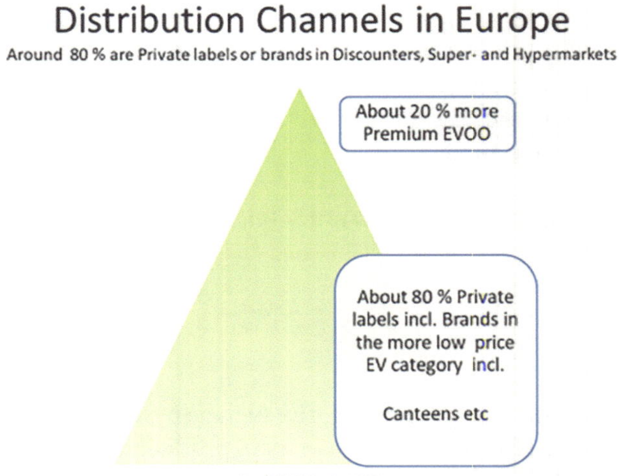

Figure 1.
80% of EVOO is sold as "private label" to supermarkets and discounter chains.

and marketing themselves. For other producers, local cooperatives handle the larger quantities of oil. In many cases, there are buyers, brokers, and sellers who sell or buy large quantities to large bottling companies in the producing countries. These producing and bottling companies work for large food chains in Europe and around the world. Up to 80% of EVOO is sold as so-called "private labels" in discount stores and large grocery chains, at least in Europe (**Figure 1**). However, where large volumes are marketed, the trade tries to buy the oils at a minimum price. Inevitably, quality can suffer as a result, especially for EVOO grades marketed as part of so-called borderline oils. These include really bad tasting and sometimes slightly defective oils, which consequently would have to be officially retested to get clarity for the classification.

1.1 The classification is something like a basic division of the total amount of olive oil

The views of market participants and consumers must be differentiated, with the question of supply and demand and the requirements for quality and price in the foreground. In the various countries, the share of olive oils in the different categories varies greatly. There are countries where on average 80% EVOO and only 20% VOO and OO are consumed. In other countries, it is even only 30% EVOO and 70% virgin olive oil, refined olive oil, and olive pomace oil. The classification is carried out by the legislator within the framework of a marketing regulation. The classification is based on the data of chemical and sensory analysis (**Figure 2**). Although the technology of oil extraction (pressing, malaxation [2], decantation), the climatic conditions, and the possibilities of chemical analysis have changed considerably in the last 30 years, there have been only minor adjustments to the legal regulations in this respect—especially with regard to the chemical control parameters. The classification is granted on the basis of the 1991 regulations in force, if the chemical analysis (officially 27 parameters) meets the requirements and the sensory test in the so-called panel test (PT) have been passed. The PT is performed by a regional, national, EC, or IOC accredited official panel of 8–12 trained olive oil tasters. It is used to verify conformity

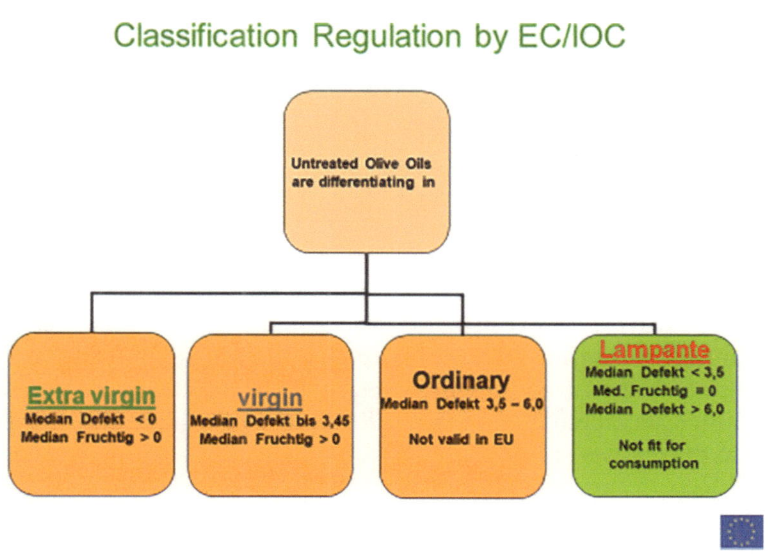

Figure 2.
IOC/EC-organoleptic criteria for olive oil classification [4].

with the requirements of Regulation (EC) 2568/91, as amended [1]. Before receiving EC approval, panels must be accredited according to EN ISO 17025 (2018) and require approval by national authorities [3] and subsequently by the European Commission. The methodology and range of requirements for sensory evaluation are also defined by the International Olive Council (IOC) [4].

Based on the panel test, the classification of the olive oil is done without prejudice to possible unauthorized material changes of the oil such as thermal posttreatment (adulteration), which cannot be detected in the panel test [5]. The aim of the panel test is the qualitative detection of negative attributes as well as positive attributes such as fruitiness (plus green-fruity or ripe-fruity), bitterness and pungency, and their quantification. The individual results of the panelists are statistically evaluated in order to obtain a sensory result that is valid and as objective as possible.

According to the currently valid criteria of the Regulation, samples that show a median for the negative attributes of "0" and also show a low fruitiness of only around 1.2 are already classified as "extra virgin." Olive oils with a fruitiness intensity of more than 5 until 7 or 8 receive the same classification. In the case of sensory defects, a distinction is also made according to the intensity of the defects. If the median for a sensory defect is between 2.5 and 3.4 in intensity, the oil still belongs to the VOO category. But if it is a rancid defect, the oil becomes the grade lampante virgin olive oil (LVOO) in a couple of weeks or months. Generally, the consumer cannot tell from the name "virgin" that this oil has a defect such as fusty or rancid. The consumer thinks more of a lower quality than of a real defect or a bad taste without being aware of the risk of spoilage. According to the current food regulations in the EC, such foodstuffs have to be labeled accordingly. In our eyes, this is a weakness of the IOC Method No. T20, No. 14/2021 or Reg. (EEC) 2568/91 in its current version, that all olive oils classified as EVOO (on an annual average more than 1.2 million tons) belong to the highest quality segment in the sense of the regulation. EVOO is not a consumer product such as pasta, milk, or potatoes; it is a food ingredient and can in many ways turn cold, hot, or warm dishes into a delicious (and nutritious) experience or even spoil in case of a sensory error. Due to its unique organoleptic properties, it requires a more objective overall assessment, including differentiation in quality that is more recognizable to the consumer, in addition to its classification.

1.2 The trade needs a reliable benchmark for determining quality in the EVOO range

An adequate approach in order to achieve more differentiation in the category of EVOO, it was found in olive oil competitions organized mainly for premium olive oils. These competitions are not subject to any official regulation but are initiatives of national associations and organizers such as the IOC for the Mario Solinas Competition [6]. In the new test sheet for this competition, the parameter harmony has to be evaluated three times (olfactive, retronasal/gustatory, and global). With all due respect to olive oil competitions, it should be noted that the IOC holds a demanding annual competition for practically less than 2% of the EVOO produced worldwide. In these kinds of competitions, the producers are awarded with gold and silver medals, certificates, and recommendations. The result is valuable top-quality oils, tasty delicacies with a high content of healthy fatty substances [7]. But because not all olives of a producer can be picked in one day and because nature does not wait for olives to ripen continuously, in most cases, only smaller quantities of these high-value oils are available to about 15–20% of the somewhat better-off consumers.

DOI: http://dx.doi.org/10.5772/intechopen.102994

These awards are only given for exceptional commitment on the part of the producers, which work with extraordinary engagement down to the last detail. Accidentally, for the remaining rest of 98% (!) of the EVOO, no qualitative distinction is foreseen although even these oils have different, more or less characteristic flavors.

Unfortunately, the consumer's knowledge of olive oil is not very common. Generally the consumer is unable to differentiate between terms such as varieties [8], cold-extracted, virgin or extra virgin, acid values. This is exploited by the trade in many countries and tolerated or supported by the legislator. Below the top qualities, there is no differentiation any more apart from mainly emotional arguments. Especially in this more affordable but very important mass market segment—which means for 80% of the consumer—a better differentiation is needed. It would help more consumers to gain confidence in a somewhat more expensive but recommendable oil.

The trade is dominated by price competition that often ignores quality aspects. For example, the 2015 IGO study [9] showed that out of 70 rather cheap samples from 15 different EU countries intentionally labeled as Extra Virgin, only 41.4% (29/70) could be confirmed as EVOO. 58.6% (41/70) had to be downgraded to the VOO category. The origin of the samples was reported as 58% EU blend, 23% Spanish blend, and 9% Italian blend.

Unlike food from certain countries such as honey or wine, olive oils from one country can be blended with oils from other countries within the EU and marketed as EU-Blends. The publication "EU-Blend" on the label is obligatory in such cases as well as all other origins from EVOO. The consumer must pay close attention to labels, including the brand name, as it is now mandatory to provide information on the label about the actual origin of the oil. Despite the wealth of information, many consumers lose their bearings and trust their personal shopping source more.

Regardless of this fact, we still assume that a not insignificant percentage of the oils sold worldwide as EVOO actually belong more to the VOO category due to sensory deficiencies. Some manufacturing companies simply refer to this type of EVOO as "borderline" EVOO. A practice that unfortunately cannot be ruled out is that olive oils with slight sensory defects are blended with extra virgin olive oils in such a way that the defects are very difficult to detect in a panel test. This means that for this type of EVOO in different panel tests (PT) different results such as VOO or EVOO can arise. The unofficial trade-declaration "borderline" depends on the requests of local stakeholders and the willingness of some suppliers, especially in those countries where the food control for olive oil is not yet so developed.

2. Material and methods

2.1 The panel test and the additional criteria harmony (QHV)

The German Olive Oil Panel (DOP), established in 1999 as a "virtual" panel [10] with the approval and support of the IOC, addressed the issue of different qualities within the EVOO category. The DOP was recognized by the IOC from 2001 to 2005 [11]. Since 2007, the DOP has been accredited according to EN ISO/IEC 17025 [3]. From 2011 to 2021, it is recognized by the German authorities, the European Commission, and the IOC [12].

The IOC has defined several guidelines and instructions for organoleptic assessment, which are continuously adapted. They include methodological aspects such as the necessary number of tasters (8–12), the basic vocabulary, the use of the profile sheet, test glasses, booth, etc. All details are described in depth in the IOC standards T20 No.14/2021 [4] and Reg. (EC) 2568/91 in their current version [1].

As mentioned in Chapter 1.2 of the introduction, the Panel test (PT) result in the European Union and third countries is only used to classify whether the oil is "extra virgin," "virgin" ("ordinary"), or "lampante." This study is concentrating on those samples only which gained the classification extra virgin.

The idea to extend the official sensory evaluation of olive oil was developed in the DOP by the panel leader Dieter Oberg already in 2003 in order to further differentiate the quality within the highest category "extra virgin." For this purpose, it was necessary to extend the sensory test and as a consequence as well the profile sheet. A further parameter "harmony" was added at the end of the sheet analogous to the testing in many olive oils competitions. The official criteria defects, fruitiness, bitterness, and pungency remain unchanged. The tester follows the Reg. of EC/IOC for the result for the classification. But then it has to be decided for one additional but very important parameter "harmony" (QHV) in the same session.

The new QHV method [13] shows a mark in the middle of the additional line on the profile sheet (**Figures 3** and **4**), which indicates the mark "5" to the taster. Around this mark is a standard quality in the sense of an average quality or the cheapest offer often found in German supermarkets and discounters. A quality without sensory defects, but with a decent average in terms of character, flavor, or pleasant persistence, for example. Only 15–30% of the EVOO in the world reach values of more than 5.5 in the harmony scale. Values of more than 7 are only achieved by qualities, which are in a range of premium products. On the left side from the value "5" between 3.5 and 4.5, one finds oils with a sensory profile, which is "not sufficient" until "bad"—depending on if a panel did not decide for a median of defect. Many of the so-called borderline oils are further left around 2.8–3.8. Oils with a recognizable sensory defect are only rated around 3 and, depending on the intensity of the defect, even lower or 0. Finally, it is the PSV who can along the results of the panel test decide if a median for a defect is reached and then will set the factor harmony to "0." In the training of tasters, the new criterion of harmony attempts to distinguish the fruitiness of an oil not only as present or absent, ripe or green. Now, for example, the differences between the olfactory impression (clean, clear, or less) and the subsequent taste impression

Intensity of perceived negative Attributes	**Intensity of perceived positive Attributes**
Fusty, muddy sediment	Fruity O green O ripe
Musty, humid, earthy	Bitter
Winey, vinegary, acid-sour	pungent
Frostbidden, wet wood	Harmony / Balance
Rancid	*Name of taster*
Others	*Code No.* *Date*

Figure 3.
Excerpt from the profile sheet with additional line for harmony evaluation.

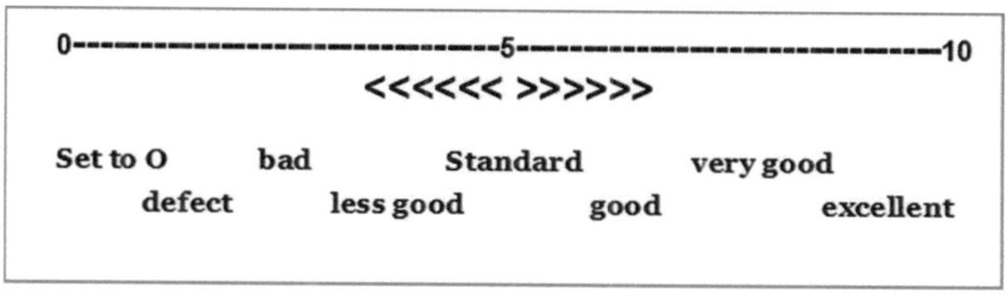

Figure 4.
Rough division of the red curve (enlarged) of ***Figure 3****.*

(harmonious or strange, unexpectedly good in the sense of the relevant criteria or not) are additionally included in the quantified result for harmony. The tasters are asked as well to recognize particular aromas in olive oils. Aromas appear to the taster as the scent of a large bouquet of flowers or as clearly recognizable individual aromas of certain leaves, fruits, vegetables, or spices. Beyond this, the persistence and mouthfeel of the sensory impression create an overall impression in the taster, which is then translated numerically. This requires a special training of the tasters, as they are not limited any more to the official criteria such as fruity-bitter-pungent but can recognize additional sensory characteristic aspects of an oil, similar to wine. There are more than 500 olive varieties [8] grown in different ways in more than 20 countries. The goal is to produce oils with a certain sensory profile, even below the premium class. Only the additional criteria of harmony can fulfill this ambition.

2.2 Methodology and practice

The official methodology for determining the QHV value includes four steps, which are described below and can be trained and performed using the feature-benefit formula in **Figure 5**. The feature-benefit diagram is a path on the way to the decision for a certain value for the taster's QHV. The taster is referred to all four additional sensory characteristics (A–D) to be evaluated, can roughly tick the differently strong or less strong expression of the individual criteria, and thus arrive at a quantified overall result, which will be sent to the PSV.

A. **Relation between olfactory impressions (odor) and gustatory impressions (taste):** The first impression of the fruitiness via nose is the basic for purity and diversity of elements, which includes already the intensity. It feeds expectations for a similar retronasal impression. The second impression via palate (retronasal) can match the impression (plus/minus) or the taster has an unexpected impression with positive or negative olfactive elements.

B. **Degree of purity and harmony/balance of the positive attributes (fruity, bitter, pungent)**: This balance increases if at the end bitterness and pungency are in a balanced relation to the fruitiness—preferably with a long persistency. Less balance means if bitterness or pungency is too dominant, and the positive impression of the fruitiness is more or less disturbed. The decision for the intensity of bitterness and/or pungency needs time.

C. **Diversity, purity, and intensity of aromas and the harmony/balance between them:** A sample can offer more positively a diversity of aromas in different intensities or just one single aroma—also in a lower intensity. A sample gives more negatively the

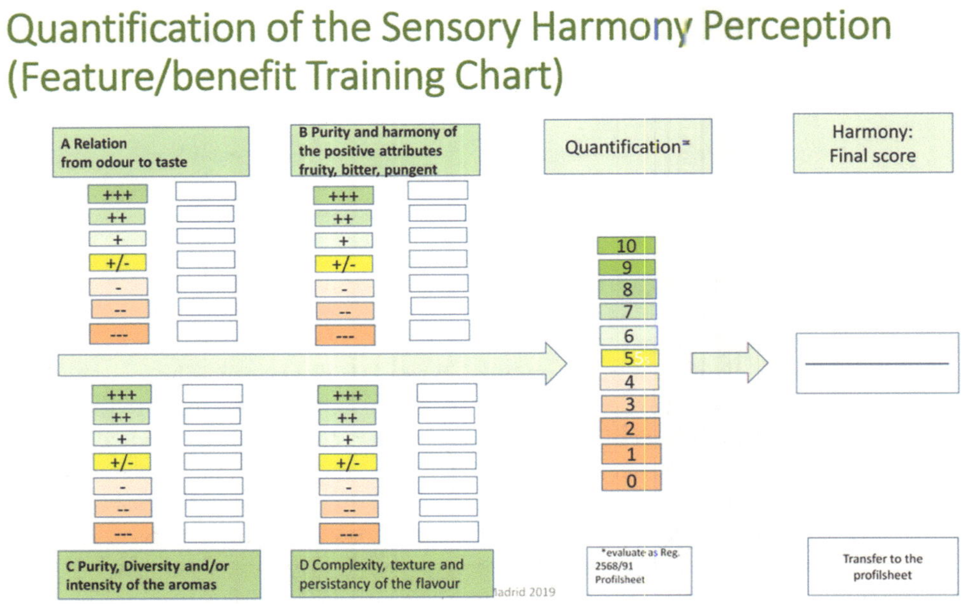

Figure 5.
Feature/benefit chart as an aid to getting started with the decision way for the additional QHV.

impression of not clearly to defined aromas and/or aromas, which have not a perfect purity or not clear to defined aromas. Some aromas can be detected retronasal only.

D. Complexity of flavor, persistency, and texture of the positive impression:
Complexity increases with the number/intensities of aromas and the flavor.
A long and pleasant persistency of the flavor is positive.
A more long texture is more positive than a short texture.

A rough description of an example: The tester decides for A with +, for B with +, for C with -, for D with ±. In the “Quantification” column of this theoretical sample, the tester first might decide mentally on the range “around 5” and then his final score 4.8. He enters this result on the profile sheet and passes it on to the PSV. Admittedly, a tester’s decision for a QHV quality “not sufficient, “upper standard,“ or “very good“ is as exact as a decision for the intensity of fruitiness or a defect. Ultimately, homogeneity in a panel for all attributes must be intensively practiced for a non-scaled line. Since so-called “outliers” can be eliminated by the PSV (see QHV methodology), the QHV evaluation with the median is as accurate as the official panel test. All the factors to be evaluated are well known to the international olive oil experts and tasters. Only the aroma designations may differ from tester to tester and from country to country, because not all aromas are common to every tester. Regardless, however, an individual aroma or aroma bouquet can be recognized by the tester in each country, as well as the evaluation of clarity and flavor. But even better, if the tasters stay longer in the panel, the continuous training and practice allow the summary of all mentioned criteria on one additional line (**Figures 3** and **4**) at the end of the official profile sheet. This means that the taster makes his evaluation in one operation first for the classification according to Reg. (EEC) 2568/91 in its current version (or the corresponding IOC standard) and then finally his decision for the value for the QHV (method DOP-2007-2).

QHV – Quantified Harmony Value	Description of the relevant EVOO Quality
Set to „O"	Valid Median of Defect
< 3,0	Defect, valid Median probably
3,1 – 4,4	Not sufficient, very bad Quality
4,5 – 5,0	Lower Standard, not sufficient
5,1 – 5,4	Upper Standard, satisfactory Quality
5,5 – 6,4	Good, high Quality
6,5 – 7,4	Very good, very high Quality
7,5 >	Excellent Quality

Figure 6.
Definition of QHV quality levels for the different grades of EVOO with the method DOP-2007-2.

The idea of this additional parameter harmony is to assign oils with the classification EVOO to additionally defined quality groups (**Figure 6**). This additional quality criterion is not regulated by law. However, the trade now has the possibility to select oils according to the requirements of the market in terms of price and quality.

The unofficial assessment of QHV in the panel test should not be mandatory, but it should be an important option for the trade. Basically, the quality of an olive oil does not remain constant, but deteriorates slightly over the course of 18 months due to decreasing fruitiness and freshness. This and the fact that in the mass market new batches have to be compiled again and again during the year due to the large quantities require constant sensory control also on the basis of specifications from the trade. These specifications must not be utopian, but must be based on realistic values depending on the season.

The QHV methodology also includes the following guidelines [13]:

- The valid median must have a robust coefficient of variation (CVr) below 10%.
- Single results that exceed from the median by 1.5 or more have to be eliminated by the PSV. In case of a panel with eight tasters, only at least six from eight tasters need to have a valid result for the median of the QHV.
- Suggestion of including the control of the Z-Score for the QHV in connection with the results of the PT for every taster.
- The PSV is as well able to moderate single results of the QHV only.
- In case a valid median of a defect is reached, the harmony value is set to zero (0) by the PSV.
- The repeatability

3. Results and discussion

3.1 Global experiences of the previous 10 years

The application of this new criterion in sensory analysis since 2011 has led to a clearly recognizable improvement in the quality of olive oils on the German-speaking market as the figures from the IGO/DOP do prove. Every year, the DOP tests about 1000 samples according to this scheme, including harmony. **Figure 7** illustrates that at the beginning in 2011, only 26% of the EVOO were rated >5.0–10 on the scale (from "higher standard" to "excellent"). By 2020, this figure had already risen to 75.6%. A measurable success for quality and thus for consumer protection. The trend from 2013 to 2020 in particular shows the remarkable way with a significant decrease in very poor qualities (QHV between 3.1 and 4.4) and also in the range of 4.5–5.0 ("lower standard"). The standard quality originally ranged from 4.5 to 5.4, but since more than 50% of all samples were in this segment, the standard quality was divided into lower (4.5–5.0) and upper standard (5.1–5.4). As a result, traders with their supporters focused only on the upper standard quality. The columns also show a remarkable increase in the range of 5.5–6.4 (QHV "good quality") for the first wave of 2017, which seems remarkable as 2017 was a difficult year in terms of harvest in different producing countries.

The QHV value is still unofficial and not part of the regulation, but it is firmly established in practice in German-speaking countries and is slowly gaining more and more acceptance among other importing and supporting companies. In the first years, the introduction process of the QHV was somewhat tough. Since 2011, the method has been presented to the professional public in further workshops and congresses [14–16] and backed up with initial figures. In countries with strong trade groups such as Germany, Austria, Switzerland, and others, the trade began to take an interest in the QHV, as media such as TV and the press—partly justified—were constantly looking for points of attack to devalue the olive oil quality at supermarket chains and discounters. The Information Office Olive Oil [17] in Germany (IGO) has therefore organized seminars at almost all relevant food companies since 2005, in which the QHV was explained as an important additional parameter. In the first years, the trade was content with at least reaching the level of >4.5 (standard quality) on

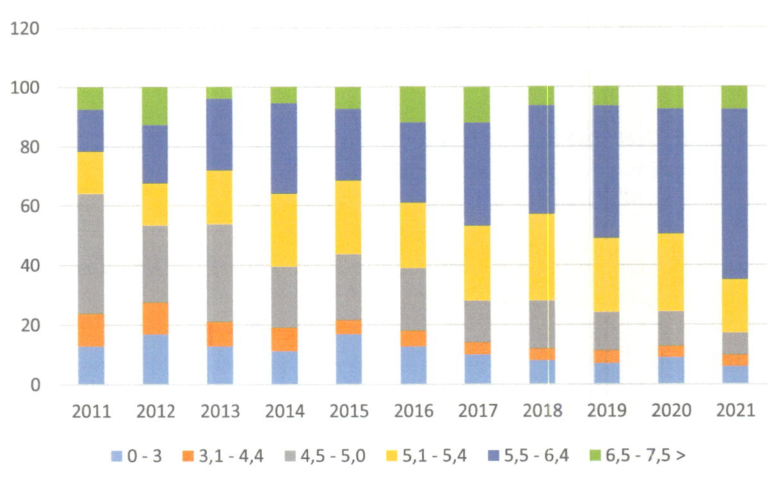

Figure 7.
QHV results 2011–.2021(%) in Germany (source: IGO/DOP), n = 7500, results 2021 preliminary).

DOI: http://dx.doi.org/10.5772/intechopen.102994

the QHV scale. After a few years, the next level followed with 5.0—5.4 (upper standard). This was certainly due to the fact that the national control medium Stiftung Warentest considered the QHV as a co-decisive criterion in their comparative tests for the quality of an olive oil. This has also led to interest in this type of assessment on the part of bottlers, traders, and producers. At the request of the trade, introductory seminars were also held in the producer countries by the IGO. This led to further quality improvements in the trade, as now the quality became measurable with the help of the QHV method, which has also been accredited by the official German inspection body (DAkkS) since 2015. Now the trade is able to demand specified sensory quality requirements (intensities of fruity, bitter, spicy plus QHV) from its suppliers. However, due to the annually fluctuating characteristics of olive oils and the natural decrease of the individual intensities, the requirements have to be redefined every year or adapted to the annual rhythm by neutral experts.

Of course, thought was given to including the QHV on the label. However, for various reasons, this was not done and the responsibility was left to the trade. On the one hand, the label for EVOO is already very extensive due to numerous legal requirements. Since the QHV would entail an additional need for explanation for the consumer and it is also only used in a few countries, this idea had to be abandoned for the time being.

But in addition the type of fruitiness—green, green/ripe, or ripe fruitiness—became also part of the EC/IOC regulation in the framework of the panel test. Our investigations showed that these differences also had an impact on the assessment of the QHV. The graphic (**Figure 8**) shows an example from Greek oils in 2018: EVOO with green flavor received QHV scores in the range of 5.6–6.4 (very good) due to aroma richness and more intense fruitiness. For oils with green and ripe taste, i.e., from fruits with advanced ripeness, most QHV scores were between 5.1 and 6.4 (upper standard to very good). Oils with a more mature taste from ripe harvested fruits mostly reached QHV values of 4.5–5.4 (standard), partly because the fruit intensity is milder and the aromas are sweeter. In the latter EVOO, the polyphenol values are already significantly lower, which, among other parameters, also affects the stability of the oil. Due to natural conditions—ripening is a continuous process—harvesting times have been moved forward somewhat in some countries to increase the proportion of green or green/ripe olives. This is also in order to gain more green aromas, a higher nutritional value, and as a consequence, a slightly higher QHV value.

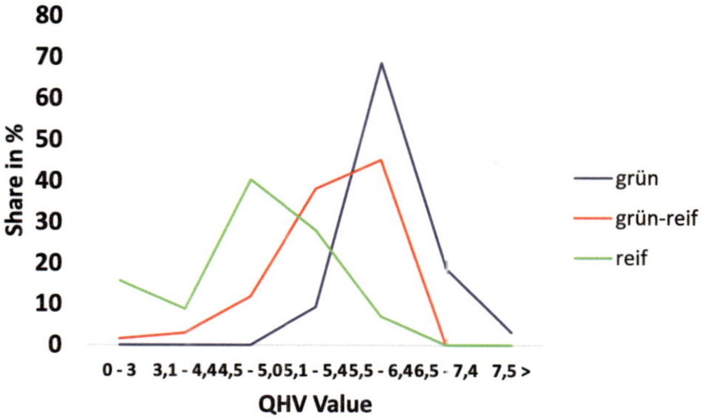

Figure 8.
Type of fruitiness and relevance for the QHV, grün = green; grün-reif = green-ripe; reif = ripe, source: 2018, n = 157 Greek samples, IGO [10].

3.2 Experiences from production countries for fruitiness and QHV

The effect of the QHV also opens the consumer's senses for special characteristics of blended and country-specific oils—and that also for more affordable oils in the so-called mass market. To this end, we have broken down certain parts of the DOP test results of the last 7 years country by country to show differences and improvements. *The reader may pay attention to the fact that the calibration of the intensity of fruitiness in the DOP panel has changed 2021 to a slightly higher level. This step was initiated and practiced by a modified method in the framework of an IOC workshop and started in year 2021, but the process for the DOP panel has not yet been finalized.

Spain is the largest producing country in the world with an average of 1.2–1.4 Mio t. But only around 50% of this quantity is officially EVOO [18]. In addition, in the 2016/17 campaign, there was a big change in the development of fruitiness in Spain (**Figures 9a** and **b**). With an earlier harvest start since 2016 for the important varieties Picual and Hojiblanca, it became possible to bottle EVOO with at least green/ripe fruitiness even after 12 months for the mass market. This resulted in average fruit intensities of 4.1–5.0 (IOC/EC method) also in the second half of the year instead of only 3.5–4.0 as in previous years. This was also expressed in a QHV value that was more than 50% in the range of 5.5–6.4 (good) from 2018 to 2020. The graphs (**Figures 9a** and **b**) also show that the significant calibration correction for fruitiness has practically no influence on the evaluation of the QHV. Most of the samples consist of different mixtures of the Picual and Hojiblanca varieties and show a stable continuity of quality thanks to skillful blending.

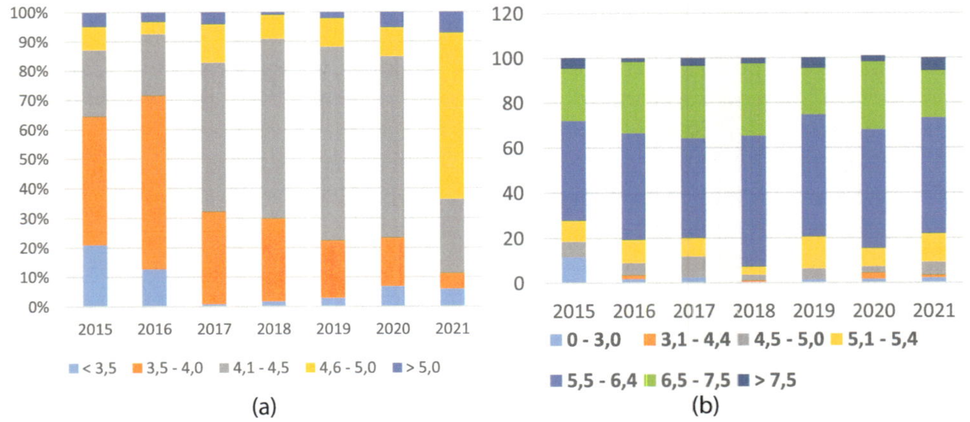

Figure 9.
(9a) Fruitiness rating Spain, 2015–2021 (%), n = 65–265, (9b) QHV rating Spain, 2015–2021 (%), n = 65–265. Source: IGO/DOP.

Italy is home to a wide range of olive varieties with very different aroma profiles. In addition to individual other varieties, Coratina, Ogliarola, Arbosana, Carolea from the southern regions of Puglia and Calabria are mainly used in rather affordable oils due to sufficient quantities. These varieties already had a high QHV level since 2015 with values between 5.5 and 7.4 (good to very good) and were able to maintain this quality over the 5 years as far as climatic conditions allowed. This also shows a very stable QHV level over the entire period (**Figures 10a** and **b**).

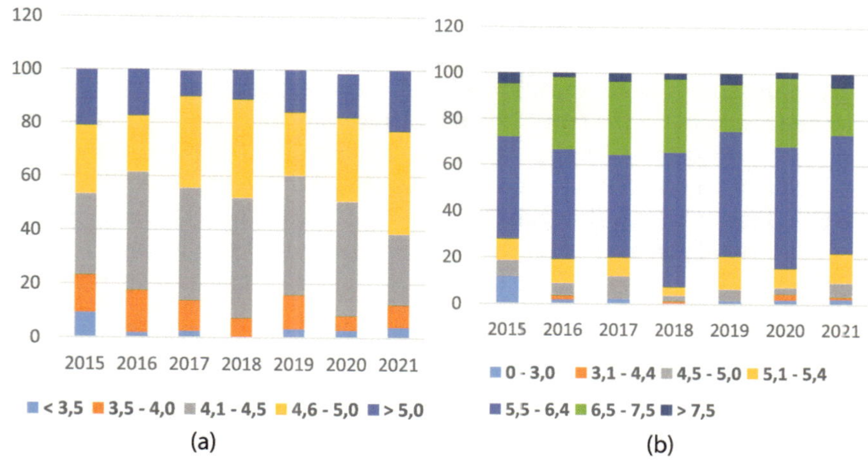

Figure 10.
(10a) Fruitiness rating Italy, −2015-2021 (%), n = 50–76, (10b) QHV rating Italy, 2015–2021, (%) n = 50–76. Source: IGO/DOP.

In contrast to the above named countries, **Greece,** the third largest producing country, is characterized by many smaller farmers. In recent years, small cooperatives have been established to pool more knowledge about the harvest, quality, and what is happening in the markets. Throughout Greece, Koroneiki is the most important variety, accounting for 70% of the fruits grown. However, the variety can grow very differently due to climatic and soil differences between the north, the Peloponnese, and the island of Crete. The more the farmers work according to a certain predetermined strategy, the more consistent and better the quality will be. It can be seen that the intensity of fruitiness has increased significantly since 2016 to around 5.0 (year 2021 is different because of the new calibration). This was accompanied by a yearly increase in QHV up to the 5.5–6.4 level (good) or at least to the 5.1–5.4 level (upper standard), while QHV scores below this level continuously decreased (**Figures 11a** and **b**).

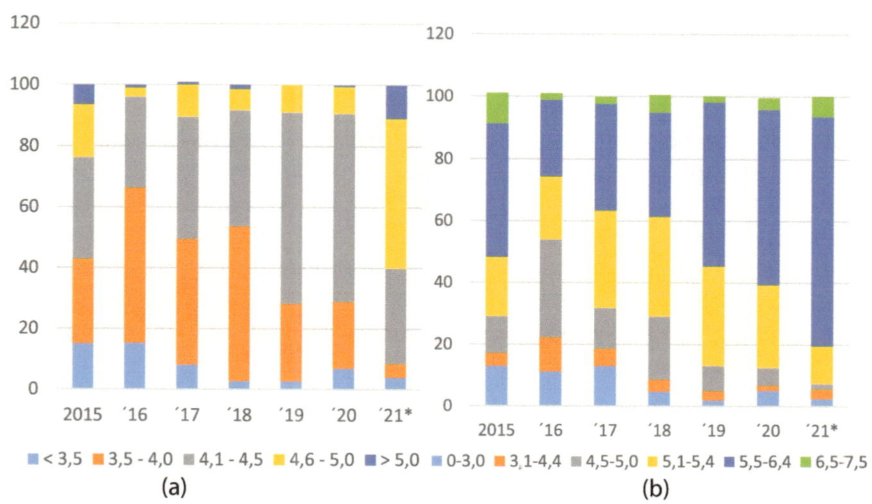

Figure 11.
(11a) Fruitiness rating Greece, 2015–2021(%) n = 95–166, (11b) QHV rating Greece, 2015–2021 (%), n = 95–166. Source: IGO/DOP.

EU Blends (EVOO from different EU producing countries, n = 130–363) also show a clear change in quality after 2015. The intensity of fruitiness (**Figure 12a**) increased continuously from 3.5–4.0 to 4.1–4.5 over the last 5 years. With the higher intensity, the character of the fruitiness also changed. Instead of predominantly ripe fruitiness as in 2015 and earlier, more oil was obtained from green to green-ripe olives, which in turn brought more intense flavors and thus also had a positive effect on the QHV (**Figure 12b**). While the QHV value 5.1–5.4 (upper standard) remained constant at a level of just over 30%, the value 5.5–6.4 (good) practically doubled from 16.2% to 35.4%.

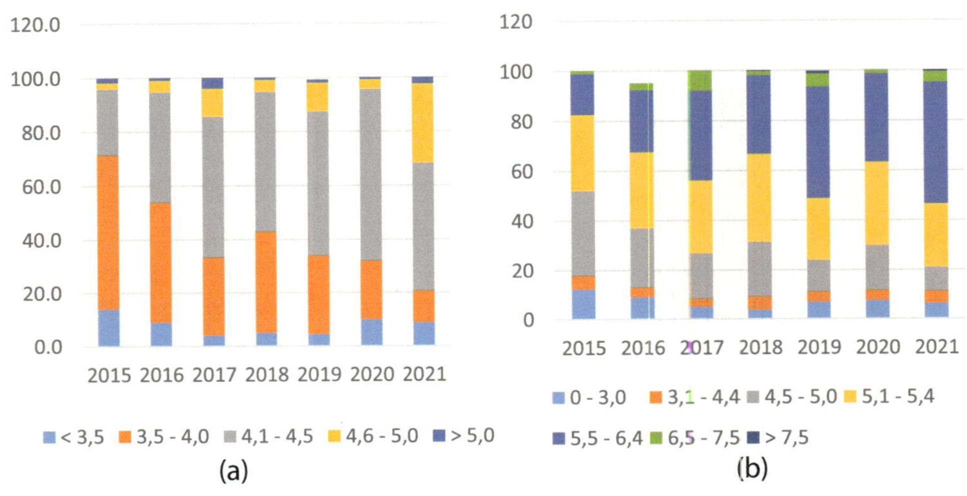

Figure 12.
(12a) Fruitiness rating EU blend 2015–2021, (%), (12b) QHV rating EU blend 2015–2021 (%). Source: IGO/DOP.

Correspondingly, the value 4.5–5.0 (lower standard) dropped from 34.1–18%. EU blends are often the "price entry quality" for many nationally active traders.

The origin from the yearly up to 1000 DOP samples is roughly 20% Greek oil, 10% Italian, 30% from Spain, and 40% EU Blend. The figures for 2021 are not yet complete. In connection with all the above results, equally excellent EVOO were tested, some of which also won prizes in international competitions. These results have not been broken down by country, but they have a negligible impact on the above results, as they only account for 5–7% of the total number of samples per year. Values above 8.5–9, even up to 10, are extremely rare. But as already mentioned: "The higher the quality, the lower the quantity available in the trade"—therefore only a few oils can reach this segment.

Samples from other countries such as Portugal, Tunisia, Turkey, Croatia, Slovenia, and Palestine were also evaluated, but the number of samples per country was not sufficient for serious statistics.

4. Conclusion

The results for the individual production countries, as well as for all samples together, undeniably show the way to a quality improvement over a long period of time that was hardly thought possible. This is true not only in German-speaking countries, where during the previous years around 1000 panel tests are conducted annually, but also for production and distribution of some international brands. Last but not least, it was the trade that took up the idea in order to get olive oil out of the headlines of the negative press. After all, olive oil was once in the top group of food products that drew attention to themselves through fraud and adulteration. But the figures on which this study is based speak primarily for Germany, Austria, and Switzerland. In these countries, the control policy of the trade often still goes far beyond the requirements of the government or the European Commission. This also applies to the sensory analysis of olive oil taste, a decisive criterion for the consumer.

With the QHV, the trade received an additional quality parameter that closes the gap between the best classification (EVOO) and measurable quality grades for this classification. The trade—in the meantime also instructed in the QHV method—works out a profile for its various olive oils with expert advisors and passes it on to the producers. These—as well QHV trained—receive thus specifications for the production or blending. This helps the trade to the same extent with the so-called "borderline problem" between the two grades EVOO and VOO. With EVOO there are good and bad ones, with VOO there are always a median of defect with intensities up to 3.5. The QHV would certify a bad EVOO a very low rating and thus inform the buyer accordingly. However, the results also show that it is possible to achieve a certain good quality level with the QHV even with supermarket and discount chains—i.e., with very large quantities. Whether to achieve "upper standard" or "very good" is up to the retailer in each country to decide. But thanks to QHV, the desired level is quantifiable.

The evidence gathered since the QHV presentation in September 2019 to the IOC Expert Group on Sensory Items held in Madrid does not indicate that the QHV method will be adopted globally. However, it may be introduced on a country-by-country basis or applied by nongovernmental bodies at the request of operators. It was found that in some olive oil producing countries, there is a kind of creeping path for improving the field of processing from agriculture via harvesting, malaxation [2] to extraction.

The QHV method was developed with a focus on the transparent presentation of the different qualities of the "extra virgin" grade of this exceptional product olive oil. Since 2011, attempts have been made to show the possibilities of the QHV to the IOC as well [13, 19] in order to start joint international trials that would enable its possible use in other countries. While the Commission in Brussels accepts or nationally tolerates some member countries' own initiatives—especially against the background of consumer protection with simultaneous proper production—the seemingly unassailable IOC seems to be pursuing its own policy under the protective shield of the UN. The national and economic benefits of olive oil are of undisputed importance in the producing countries. But it should also be transparent in the importing countries and compatible with the qualitative and legal requirements of consumers in their countries.

The QHV provides security in the assessment of EVOO qualities, it creates measurable transparency in the field of marginal oils, it limits the possibilities of fraud and helps in the detection of adulteration. With the QHV, there are fewer complaints and more satisfied customers.

Abbreviations

QHV	Quantified Harmony Value
EVOO	Extra Virgin Olive Oil
VOO	Virgin Olive Oil

LVOO	Lampante Virgin Olive Oil
OO	Olive Oil (Refined Olive Oil Plus EVOO)
PT	Panel Test
IOC	International Olive Council
DOP	German Olive Oil Panel
PSV	Panel Supervisor/Panel Leader

Author details

Dietrich Oberg
Association for Care and Support of the Value of Extra Virgin Olive Oil e.V./Information Office Olive Oil (IGO), Wessling, Germany

*Address all correspondence to: d.oberg@olivenoel-info.de

References

[1] Regulation (EWG) 2568/91 in its actual version on the Characteristics of Olive Oil and Olive Residue Oil and the relevant Method of Analysis. Official Journal of the EEC. 1991;L **248**:01-83. ISSN: 0376-9453

[2] Clodoveo M. Department of Agro-Environmental and Territorial Sciences, Agricultural Faculty, University of Bari, ItalyMalaxation: Influence on virgin Olive Oil Quality. Past, present and future. An Overview. Trends in Food Science & Technology. 2012;**25**:13e22 Amendment with Regulation (EU) 2015/1833 of October 12, 2015 on the Characteristics of Olive Oil and Olive Residue Oil and the Relevant Method of Analysis

[3] German Accreditation Office (DAkkS), Berlin. DIN ISO 17025General Requirements for the Competence of Testing and Calibration Laboratories. Berlin: DAkkS; 2018

[4] International Olive Council (IOC), Madrid. General methodology for the organoleptic assessment of virgin Olive Oil. Madrid: COI/T.20/Doc. No. 15/Rev 6/; 2019

[5] International Olive Council (IOC), Madrid. Sensory Analysis of Olive Oil– Method for the organoleptic assessment of virgin Olive Oil. Madrid: COI/T.28/ Doc.No. 1/Rev 5; 2019

[6] International Olive Council (IOC), Madrid. Rules for the international Extra Virgin Olive Oil Competition Mario Solinas Award. Madrid: DEC-III.5/114-VI/; 2021

[7] Delgado C, Guinard J-X. University of California, Dep. Of Food Science and Technologogy, Davis (California)Internal and external Quality Mapping as a new approach to the Evaluating of Sensory Quality. Technology Journal of Sensory Studies. © 2012 Wiley Periodicals Inc. 2012;**27**:332-343

[8] International Olive Council (IOC), Madrid. World Catalogue of Olive Varieties. Madrid: International Olive Council (IOC); 2000. p. 95

[9] Oberg D, Association for Care and Support of the Value of Extra Virgin Olive Oil (IGO), Wessling (Germany). Influence of the Harmony Value for Reliable Sensory Data in Private Label Quality of EVOO in the EU. Florence: IGO Study, presented at Euro Fed. Lipid Congress; 2015 and DG Fat Congress Nuremberg (2019)

[10] Bongartz A, Popp M, (University of Applied Siences, Wädenswil (Switzreland); Retsch R, Panelsupervisor, Nuremberg (Germany), Data Validity within the Sensory Evaluation of EVOO, in situ (central) vs. remote (decentral), DG Fett Congress, Leipzig/Germany, 2021

[11] International Olive Council (IOC), Madrid. Certification for the Sensory Testing LaboratoryGerman Olive Oil Panel (DOP). Madrid: IOC; 2002-2005

[12] International Olive Council (IOC), Madrid, Certification for the Sensory Testing Laboratory "German Olive Oil Panel", 2011 and subsequent years until 2021; 2021

[13] Bongartz A, University of Applied Sciences (ZHAW), Wädenswil (Switzerland); Oberg D, Lecturer at ZHAW, InTech Book ; 2016, "Products from Olive Tree", 138-155, Chapter 8,

[14] Oberg D, Information Office Olive Oil, Wessling (Germany). Benefits from

an extended sensory Assessment for EVOO. Munich/Germany: Presented at Euro. Fed. Lipid Congress; 2010

[15] Oberg D, Lecturer, Panelsupervisor, Wessling (Germany); Bongartz A, (University of Applied Sciences, Wädenswil (Switzerland), "Influence of the Harmony Value on Quality Discrimination in the Grade of EVOO". Presented at Euro. Fed. Lipid Congress. Antalya/Turkey, 2013

[16] Bongartz A, University of Applied Sciences, Wädenswil (Switzerland), Oberg D, Lecturer, former Panelsupervisor, Wessling (Germany) "Quantified Harmony Value–The Differentiation Parameter for the Grade EVOO", Poster session Euro. Fed. Lipid Congress. Seville, 2019

[17] IGO (Information Office Olive Oil e. V., Wessling (Germany)identical with "Association for Care and Support for the Value of EVOO e. V". Germany: IGO; founded 1998

[18] Teatro Naturale, Source "Centro Tecnologico Agrialimentario di Extremadura (CTAEX), "The Truth about Olive Oil in Spain", 2020

[19] Bongartz A. University for Applied Sciences, Wädenswil (Switzerland), Oberg D, Lecturer, Sensory Evaluation of EVOO extended to include the factor Harmony. Journal of Agricular Science and Technology A. 2011;**1**:422-435

Chapter 9

The Effect of Olive Leaf Extract on Systolic and Diastolic Blood Pressure in Adults: A Systemic Review and Meta-Analysis

Fatemeh Rahimianfar

Abstract

Hypertension (HTN) is one of the most common disorders and increases the risk of cardiovascular diseases (CVD), which are one of the main causes of death in the world. The Mediterranean diet has the efficacy to modulate CVD risk factors such as HTN, mainly because of olive tree products, which are the most pivotal ingredients in this diet. Among the olive tree products, olive leaf consists of many sorts of phenolic compounds and has several beneficial effects on human body, such as antioxidant, anti-atherosclerotic, anti-inflammatory and especially anti-hypertensive effects. So, we conducted a new systematic review and meta-analysis on anti-hypertensive effect of OLE in adults. The meta-analysis showed a significant reduction effect of OLE on systolic blood pressure. The anti-hypertensive effect of OLE is mainly considered due to its principal phenolic compound known as oleuropein (OL), which reduces blood pressure by a number of particular mechanisms associated with its specific chemical characteristics.

Keywords: olive leaf, oleuropein, anti-hypertensive, phenolic compound, blood pressure

1. Introduction

CVD is one of the most common causes of death in the world [1]. Some disorders such as HTN, type 2 diabetes mellitus (DM2), hypercholesterolemia, atherosclerosis and inflammatory disorders can increase the risk of CVD [2]. Among these disorders, HTN is one of the most common diseases imposed by modern lifestyle in terms of decreased physical activity and unbalanced lipid-rich diet [3].

It is estimated that around 30% of the world population will get involved with HTN by 2025 [4]. HTN gradually develops without notice, hence possibly aggravating such fatal diseases as CVD and chronic heart failure (CHF) [3]. There are several risk factors for HTN, such as family history, genetic and environmental factors [4]. The prevalence in females is dependent on age. In other words, prevalence of HTN in women >50 years old strongly increases. For instance, high blood pressure ratio in women compared with men increments from 0.6 to 0.7 at the age of 30 years old, reaching 1.1 to 1.2 at the age of 65 years old [5]. CVD risks augment throughout the blood pressure range, which begins at 115/75 mmHg. The blood pressure which is higher than 140/90 mmHg needs intervention [3].

Drugs decreasing blood pressure including angiotensin-converting enzyme (ACE) inhibitors, angiotensin-receptor blockers (ARBs), calcium channel blockers (CCBs), α-blockers and diuretics dwindle the complications of HTN [6]. However, most patients suffer from enhanced adverse drug reactions even in common doses and medication costs because of needing ≥2 drugs to attain blood pressure goals (< 140/90 mmHg or <130/80 mmHg in DM or chronic kidney disease) [6]. Whilst chemical drugs are necessary for treating and dominating the cardiovascular risk factors mentioned in the previous lines, diet also plays an important role in modulating them.

The Mediterranean diet is one of the most supreme in the world in terms of preventing chronic illnesses, such as CVD [2, 7, 8]. The bulk of the Mediterranean diet originates from plant sources of which olive tree products are the quintessential ingredient [9].

Olive tree (Olea europaea) belongs to genus Olea of the Oleaceae family [10]. The parts used in olive tree are leaf, fruit and skin. In ancient times, people applied olive tree, particularly olive leaves to treat fever, gout, wounds, diabetes, atherosclerosis and HTN [3, 11]. As a matter of fact, the leaf of the olive tree has several beneficial effects on human health attributed, in part, to hypocholesterolemic, antioxidant, antimicrobial, hypoglycaemic, anti-inflammatory, anti-atherosclerotic, and especially anti-hypertensive effects [1, 7, 12]. The uses of olive leaf for humans are abundant. However, our aim is to focus on the anti-hypertensive yield.

2. Anti-hypertensive effect of olive leaf extract

There are lots of animals and human trials conducted to inquire about the anti-hypertensive effect of OLE. The animal studies are mostly conducted on rats. In 2002, Khayyal et al. performed an 8-week investigation into the effects of oral administration of OLE at different levels (25, 50 or 100 mg/kg/day) on blood pressure in rats rendered hypertensive by oral doses of 4-week L-NAME (NG-nitro-L-arginine methyl ester, 50 mg/kg/day). They reported a dose-dependent prophylactic influence against the ascent in blood pressure induced by L-NAME and the greatest effect was related to the 100 mg/kg of the extract [13]. Besides these observations, the effect of OLE has also been researched in 2016 by Romero et al. In this study, a 5-week investigation on OLE (15% w/w OL) in spontaneously hypertensive rats at 30 mg/kg body weight reported a significant reduction of systolic blood pressure (−21.6 ± 5.5 mmHg) [14]. Although there are many other animal studies in this field, we intend to talk more about human clinical trials. Here, we collected a systematic review on a number of human randomised controlled trials (RCTs), which have been conducted to investigate the effect of OLE compared to placebo on systolic and diastolic blood pressure as primary or secondary outcomes in adults.

In 2008, an open, controlled, parallel-group, co-twin trial was carried out for 8 weeks on the anti-hypertensive effect of OLE in 40-borderline hypertensive monozygotic twins (age: 16–60) by Perrinjaquet-Moccetti et al. There were two parallel experiments, the first being the effect of a 500 mg OLE tablet (equivalent to 104 mg OL) once daily compared with no medication. The second experiment involved a 500 mg OLE tablet once daily compared with that of 1000 mg (equivalent to 208 mg OL) divided into two distinct doses. As a result, they revealed a significant dose-dependent decrease in blood pressure within pairs, with mean systolic differences of ≤6 mmHg (500 mg vs control) and ≤13 mmHg (1000 vs 500 mg), and diastolic differences of ≤5 mmHg. Also, mean blood pressure had significantly decreased just for the high dose group [3].

Elsewhere, in 2008, Saberi et al. verified the anti-hypertensive effect of OLE in another way. They enrolled 64 mild to moderate hypertensive patients with normal treatment resistance. They randomised the patients into two distinct groups (n=32 intervention & n=32 placebo). In the intervention group, each person received a 1000 mg OLE capsule divided into three doses daily. Consequently, the study demonstrated a significant decrease in mean systolic blood pressure. They found no remarkable effect on diastolic blood pressure and mean arterial pressure in the intervention group compared to before OLE treatment, despite the fact that there was meaningful diastolic blood pressure reduction in the OLE group compared to the placebo group [5].

In contrast, De Bock et al. who performed a 12-week randomized, placebo-controlled, crossover trial in 2013, demonstrated a different result. They assessed the effect of an OLE capsule (51.1 mg oleuropein; 9.7 mg hydroxytyrosol) daily on cardiovascular risk factors and insulin action in middle-aged overweight men. 46 participants (aged 35–55 years; BMI 25–30 kg/m2) randomly consumed OLE or placebo for 12 weeks with crossing over to the alternate arm after a 6-week washout. As a result, there were no remarkable changes in ambulatory (24-hour) blood pressure [15].

Further, in 2017, Lockyer et al. performed a randomised, double-blind, controlled, crossover trial to investigate the influence of a liquid-form OLE (136mg oleuropein; 6mg hydroxytyrosol) on blood pressure in prehypertensive patients. They used ambulatory blood pressure as the primary endpoint. The participants were 60 male subjects aged 45.3 ± 12.7, and body mass index (BMI) 27.0 ± 3.4 kg/m^2. They received either OLE or placebo for 6 weeks before switching to the other treatment after a 4-week washout. After tracking down the 24-hour and daytime blood pressure in patients, they represented a marked daytime and 24-hour systolic and diastolic blood pressure reduction (about 3 mmHg) compared to the control group (placebo) [7].

In 2020, Yaghoobzadeh et al. performed a 12-week investigation into the effect of OLE on cardiometabolic profiles in patients (aged 30–60) with essential HTN. The trial participants were randomly selected regarding intervention and control groups. (n= 30 intervention & n=30 placebo). As a result, a 250 mg OLE capsule, twice daily, could decrease systolic blood pressure significantly. However, it did not show a meaningful effect on the diastolic part [4].

As you notice, all of these studies show the anti-hypertensive effect of OLE, except the experiment conducted by De Bock et al. [15]. The most important differences between this study and other studies might be related to the study design, type of disease, nature of OLE, duration of extract in-take, patients' compliance and inclusion/exclusion criteria [4]. There is a systemic review and meta-analysis conducted by Muhammad Asyraf Ismail et al. in 2021 to show the effect of OLE on cardiometabolic profile in prehypertensive and hypertensive adults [1]. However, with all due respect to the authors of this study, there were a number of cases that encouraged us to make some updates with more accurate results. To clarify, the previous meta-analysis included 5 trials. Among these trials, in 2019, Javadi et al. did not investigate the effect of OLE on blood pressure [16]. Wong et al., studied a combined extract [17]. So, this study is not able to show us the pure effect of OLE. Susalit et al., compared OLE effect with a very strong anti-hypertensive drug (captopril) [6]. These three trials made the previous meta-analysis non-accurate, and they also excluded three useful RCTs for different reasons. De Bock et al. [15] were deleted because of not involving prehypertensive or hypertensive group in the study. Saberi et al. [5], was also deleted due to being a non-English RCT in addition to Lockyer et al. [7], who could not retrieve data after contacting the author [1]. Ultimately, the previous meta-analysis could not demonstrate the accurate effect of OLE on blood pressure. Hence, to determine the OLE effect on systolic and diastolic blood pressure, we aimed to perform a meta-analysis of these 5 human trials (**Figures 1** and **2**).

Our meta-analysis shows that OLE has a significant effect on the reduction of systolic blood pressure. But, its effect on diastolic blood pressure is not meaningful. The anti-hypertensive property of the olive leaf is due to its phenolic compounds.

2.1 Phenolic compounds

Phenolic compounds are assorted as secondary metabolites that have a restricted distribution without any explicit function in general metabolism [10]. On the other hand, primary metabolites including nucleic acid, carbohydrate, protein, lipid and cofactors, are involved in the synthesis of substances that are pivotal for the growth of all organisms [18]. Olive tree polyphenols are present in the plant to combat pathogens inducing bacterial infections and to react to pests and insect injuries [19, 20]. There are a wide variety of phenolic compounds in Olea europaea and its by-products with much more concentration in olive leaves (comparison, 145 mg total phenolics/100 g fresh leaf compared to 110 mg/100 g olive fruit and 23 mg/100 ml extra virgin olive oil) [1, 7, 15]. Another comparison confirms the much more concentration of total polyphenols in olive leaves is relative to the olive oil and fruit; 1350 mg/kg fresh olive leaf versus 232 ± 15 mg/kg of extra virgin olive oil [21, 22]. High content of phenolic compounds in olive leaf excited the interest of many scholars to continue the investigations with animals and humans, and that resulted in realizing the beneficial health effects such as anti-hypertensive effects [23]. Major phenolic compounds extracted from olive leaf are categorised in the following.

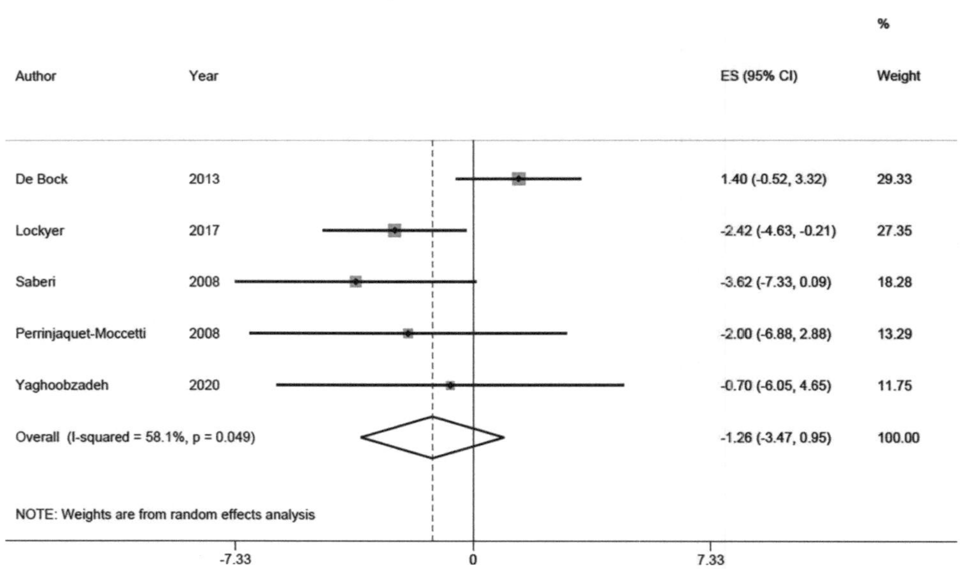

Figure 1.
The meta-analysis of OLE compared to placebo or no treatment. Outcome: diastolic blood pressure (mmHg).

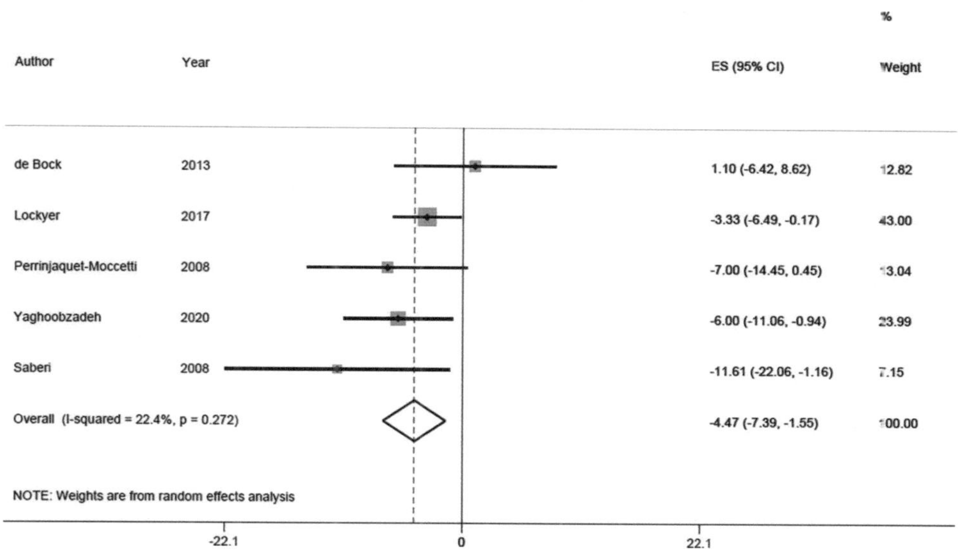

Figure 2.
The met-analysis of OLE compared to placebo or no treatment. Outcome: systolic blood pressure (mmHg).

2.1.1 Olive leaf phenolic compounds categorisation

Some researchers categorised the phenolic compounds of olive tree in 5 groups: flavones (apigenin-7-glucoside, diosmetin, diosmetin-7-glucoside, luteolin and luteolin-7-glucoside); flavonols (rutin); flavan-3-ols; oleuropeosides (verbascoside and OL) and substituted phenols (vanillin, vanillic acid, caffeic acid, tyrosol and hydroxytyrosol) [24]. Also, some researchers categorised the phenolic compounds of olive leaves into three distinct groups: (1) phenolic acids like vanillic acid, syringic acid, salicylic acid, vanillin, etc. (2) Flavonoids like luteolin, rutin, and apigenin-7-o-glucoside, luteolin-7-o-glucoside, etc. (3) Hydroxycinnamates and structurally related compounds like verbascoside, oleoside, ligasterol, oleuropein, etc. [25]. The most abundant phenolic compound identified in olive leaves is oleuropein, followed by hydroxytyrosol, luteolin-7-glucosides, verbascoside, and apigenin-7-glucosides [23]. It has been demonstrated that there are some factors that affect the chemical composition variability of olive leaves, like origin, storage conditions, proportion of branches existing in the extract, weather conditions, moisture content and degree of soil contamination [26, 27]. On the other hand, some processes such as drying and extraction enable us to change nutritional composition of the OLE [28]. Oleuropein, the principal phenolic compound in olive leaf has a significant impact on the reduction of blood pressure due to the potential mechanisms of action with its specific chemical characteristics [2].

2.1.1.1 Oleuropein

Oleuropein (OL) is a glycosylated secoiridoid that uniquely exists in plants of the Oleaceae family, presented in olive leaves at higher concentrations, and representing 1–14% of olive leaf weight, includes oleuropein in contrast with 0.005–0.12 % of olive oil weight [25, 29, 30]. OL is also known as a coumarine-like compound presented in olive trees [8]. It is an elenolic ester of hydroxytyrosol (HT), in addition to an oleosidic skeleton possessed in common to the secoiridoid glucosides of Oleaceae

[11]. In fact, HT, known as 2-(3,4-Di-hydroxyphenyl)-ethanol is the precursor of OL and the major phenolic compound in extra virgin olive oil [25, 31]. The chemical formula of one oleuropein molecule is $C_{25}H_{32}O_{13}$ with molar mass equals to 540.518 $g.mol^{-1}$ in its standard state (at 25°C [77°F], 100 kPa) (**Figure 3**) [32]. OL has been distinguished in olive flesh, leaf, seed and peel of green (unripe) olive and is an active substance of olive leaves. Its concentration declines during maturation phase happening in olive fruits because of undergoing hydrolysis yielding different products, such as HT [8, 33, 34]. It creates the bitter taste of olive that must be removed by immersion in lye, hence generating an edible olive, known as table olive [29]. OL content in olive leaves varies depending on the cultivar, production area and leaf tissue conditions (frozen, dried or fresh) [11]. There is the possibility of extracting OL molecules by some special methods explained in the following.

2.1.1.2 Extracting methods

There are various extracting methods of phenolic compounds from olive leaves (after drying and milling), including solid-liquid extraction by maceration and soxhlet extraction utilizing water-methanol blends or hexane to yield OLE [1, 35, 36]. For more explanation on one of the most common techniques, mixing the specific quantity of dried olive leaf powder with an aqueous alcohol solution, incubating there to produce an alcohol extract. after a draining process, the crude extract will be dried again and treated with alcohol and water solution at least two more times. Then, by distilling the mixed extract under vacuum, the OLE will be produced [6, 12]. OL can be chemically decomposed into two different products, including hydroxytyrosol (HT) and elenolic acid by distinct factors such as high temperature, acid, base, light,

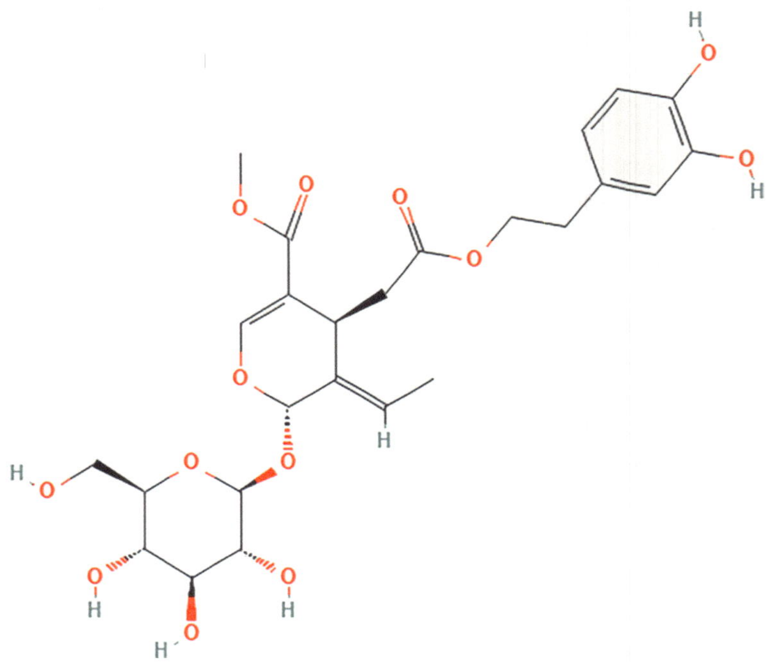

Figure 3.
Oleuropein chemical structure. Source: pubChem. URL: https://pubchem.ncbi.nlm.nih.gov/compound/5281544. Description: data deposited in or computed by pubChem.

metal ions, etc. [37]. This process assembles the enzymatic hydrolysis of this phenolic compound that occurs in human body. However, the studies conducted to exactly specify what happens to this phenolic compound extracted from olive leaves during absorption from small intestine and colon to the blood circulation, have mentioned scattered findings. Therefore, we go directly to the mechanism of its action in the body.

2.1.1.3 Mechanisms of action

The studies performed in human models to show the mechanisms of action for anti-hypertensive property of OL are scarce and have been conducted much more in vitro. This mysterious compound is endowed with anti-hypertensive property which is thought to be due to its influence on membrane receptors and/or enzymes involved in cell signalling, including ACE, L-type Ca^{2+} channels, nitric oxide (NO) and reactive oxygen species (ROS), or to clarify, the metabolite of OL inhibits ACE. Another mechanism is that a degraded product of OL (3,4-dihydroxy-phenyl-ethanol) directly affects L-type Ca^{2+} channels as an antagonist resulting in blocking the channels [29, 38, 39]. In fact, OL has synergic effects with other active substances in OLE to present ACEI and CCB activity in the body. Also, the Vasodilator effect of OLE justifies its anti-hypertensive activity [1, 29]. This phenolic compound performs a particular task to increase NO bioavailability and expression of the inducible form of endothelial NO synthase (e NOS) studied in animal subjects [9, 25, 29, 40]. As a matter of fact, OL reacts with NO and its noxious derivative peroxynitrite (−OONO). There is a possibility that OL increases NO production via modifying two specific enzymes: nicotinamide adenine dinucleotide phosphate-oxidase (NADPH-oxidase) and NO synthase [41]. These mechanisms modulate NO bioavailability, thus improving vascular function and ultimately reducing blood pressure [25]. The last one influences ROS. ROS play a significant role in the development of oxidative stress, which also encompasses the principal role in the pathology of HTN. ROS are produced permanently in the human body. They are indispensable for several mechanisms happening in the cells, such as chemical signalling, immune performance and energy production [24]. When the balance between ROS and antioxidants upsets, meaning ROS level more than the other, the cell makes oxidative stress [2, 9]. Indeed, an excess in the production of ROS which could be controlled by a number of enzymes, including glutathione peroxidase, catalase (CAT) and superoxide dismutase (SOD), enable to damage lipids, proteins and DNA in the cells particularly cardiovascular cells, are even able to ruin the vascular function and structure [2, 42, 43]. So oxidative injury increases the risk of CVD. In this regard, the OL molecule consists of some active components that have determined scaveng-ing functions [44]. So, there is a potential antioxidant property that is suggested to be related to the H-atom donation from the OL phenolic groups [8, 33, 45]. In other words, OL preserves paraventricular nucleus (PVN) of the hypothalamus from oxidative stress. OL activates the Nuclear factor erythroid 2-related factor 2 (Nrf2)-mediated signalling pathway and finally, it improves mitochondrial function. Thus provides an exquisite way to treat HTN [1, 8, 46]. Hence, antioxidant property of OLE enhances its antihypertensive yield.

2.2 OLE safety

In spite of the beneficial health properties of OLE in human body, it is essential to be determined what dosage of this extract will be safe for the body. Many studies aimed at this indicated acute OLE toxicity (2000 mg/kg) and also 4-week OLE toxicity (100–400 mg/kg) revealed no symptoms of toxicity in subjects [47, 48]. However, another study reported the opposite result by noticing bleeding in the liver and kidneys of rats when using OLE [49].

3. Conclusions

OLE can reduce both systolic and diastolic blood pressure in human body. The effect of OLE on systolic blood pressure is more significant and mostly depends on the dose of the extract for diastolic part. The anti-hypertensive effect of OLE is mostly due to OL. The two most common methods of OL extraction are maceration and soxhlet. There are special mechanisms with which OL reduces blood pressure. For instance, inhibiting ACE, blocking L-type Ca^{2+} channels, possessing vasodilator activity by increasing NO bioavailability and having anti-oxidant properties.

Acknowledgements

Fatemeh Rahimianfar would like to appreciate deeply the following people who contributed to completing the full chapter: Dr. Sepideh Soltani and Dr. Mahdieh Sadat Mousavi Rad helped to perform the meta-analysis.

Conflict of interest

The author declares no conflict of interest.

Author details

Fatemeh Rahimianfar
Shahid Sadoughi University of Medical Sciences and Health Services, Yazd, Iran

*Address all correspondence to: rahimianfarf@gamil.com

References

[1] Ismail MA, Norhayati MN, Mohamad N. Olive leaf extract effect on cardiometabolic profile among adults with prehypertension and hypertension: A systematic review and meta-analysis. Peer Journal. 2021;**9**:e11173. DOI: 10.7717/peerj.11173

[2] Mehmood A et al. A review on management of cardiovascular diseases by olive polyphenols. Food Science & Nutrition. 2020;**8**(9):4639-4655. DOI: 10.1002/fsn3.1668

[3] Perrinjaquet-Moccetti T et al. Food supplementation with an olive (Olea europaea L.) leaf extract reduces blood pressure in borderline hypertensive monozygotic twins. Phytotherapy Research. 2008;**22**(9):1239-1242. DOI: 10.1002/ptr.2455

[4] Yaghoobzadeh H, Mehravar S, Javadi H, Memarzadeh MR, Mirhashemi SM. Determining cardiometabolic and antioxidant effects of olive leaf extract in patients with essential hypertension. The Journal of Qazvin University of Medical Sciences. 2020;**23**(5):372-381. DOI: 10.32598/jqums.23.5.372

[5] Saberi M, Kazemisaleh D, Bolurian V. Effect of olive leaf on mild to moderate hypertension resistant to normal treatments. Journal of Medicinal Plants. 2008;**7**(27):52-59

[6] Susalit E et al. Olive (Olea europaea) leaf extract effective in patients with stage-1 hypertension: Comparison with Captopril. Phytomedicine. 2011;**18**(4):251-258. DOI: 10.1016/j.phymed.2010.08.016

[7] Lockyer S et al. Impact of phenolic-rich olive leaf extract on blood pressure, plasma lipids and inflammatory markers: A randomised controlled trial. European Journal of Nutrition. 2017;**56**(4):1421-1432. DOI: 10.1007/s00394-016-1188-y

[8] Castejón ML et al. Potential protective role exerted by secoiridoids from Olea europaea L. in cancer, cardiovascular, neurodegenerative, aging-related, and immunoinflammatory diseases. Antioxidants. 2020;**9**(2):149. DOI: 10.3390/antiox9020149

[9] Hermans MP et al. Supplementation effect of a combination of olive (Olea europea L.) leaf and fruit extracts in the clinical management of hypertension and metabolic syndrome. Antioxidants. 2020;**9**(9):872. DOI: 10.3390/antiox9090872

[10] Ryan D et al. Biotransformations of phenolic compounds in Olea europaea L. Scientia Horticulturae. 2002;**92**(2):147-176

[11] Nediani C et al. Oleuropein, a bioactive compound from Olea europaea L., as a potential preventive and therapeutic agent in non-communicable diseases. Antioxidants. 2019;**8**(12):578. DOI: 10.3390/antiox8120578

[12] Breakspear I, Guillaume C. A quantitative phytochemical comparison of olive leaf extracts on the Australian market. Molecules. 2020;**25**(18):4099. DOI: 10.3390/molecules25184099

[13] Khayyal MT et al. Blood pressure lowering effect of an olive leaf extract (Olea Europaed) in L-NAME induced hypertension in rats. Arzneimittel-Forschung. 2002;**52**(11):797-802. DOI: 10.1055/s-0031-1299970

[14] Romero M et al. Antihypertensive effects of oleuropein-enriched olive leaf

extract in spontaneously hypertensive rats. Food & Function. 2016;7(1):584-593. DOI: 10.1039/c5fo01101a

[15] de Bock M et al. Olive (Olea europaea L.) leaf polyphenols improve insulin sensitivity in middle-aged overweight men: A randomized, placebo-controlled, crossover trial. PLoS One. 2013;**8**(3):e57622. DOI: 10.1371/journal.pone.0057622

[16] Javadi H et al. Effects of olive leaf extract on metabolic response, liver and kidney functions and inflammatory biomarkers in hypertensive patients. Pakistan Journal of Biological Sciences. 2019;**22**(7):342-348. DOI: 10.3923/pjbs.2019.342.348

[17] Wong RHX et al. Antihypertensive potential of combined extracts of olive leaf, green coffee bean and beetroot: A randomized, double-blind. Placebo-Controlled Crossover Trial. Nutrients. 2014;**6**(11):4881-4894. DOI: 10.3390/nu6114881

[18] Haslam E. Shikimic Acid: Metabolism and Metabolites. Hoboken, NJ: Wiley; 1993

[19] Pandey KB, Rizvi SI. Plant polyphenols as dietary antioxidants in human health and disease. Oxidative Medicine and Cellular Longevity. 2009;**2**(5):270-278. DOI: 10.4161/oxim.2.5.9498

[20] Japón-Luján R, Luque-Rodríguez J, De Castro ML. Dynamic ultrasound-assisted extraction of oleuropein and related biophenols from olive leaves. Journal of Chromatography A. 2006;**1108**(1):76-82

[21] Carnevale R et al. Extra virgin olive oil improves post-prandial glycemic and lipid profile in patients with impaired fasting glucose. Clinical Nutrition. 2017;**36**(3):782-787. DOI: 10.1016/j.clnu.2016.05.016

[22] Ghanbari R et al. Valuable nutrients and functional bioactives in different parts of olive (Olea europaea L.)—A review. International Journal of Molecular Sciences. 2012;**13**(3):3291-3340. DOI: 10.3390/ijms13033291

[23] Vogel P et al. Polyphenols benefits of olive leaf (Olea europaea l) to human health. Nutrición Hospitalaria. 2015;**31**(3):1427-1433. DOI: 10.3305/nh.2015.31.3.8400

[24] Benavente-Garcıa O et al. Antioxidant activity of phenolics extracted from Olea europaea L. leaves. Food Chemistry. 2000;**68**(4):457-462

[25] Lockyer S et al. Olive leaf phenolics and cardiovascular risk reduction: Physiological effects and mechanisms of action. Nutrition and Aging. 2012;**1**(2):125-140. DOI: 10.3233/NUA-2012-0011

[26] Martín-García A, Molina-Alcaide E. Effect of different drying procedures on the nutritive value of olive (Olea europaea var. europaea) leaves for ruminants. Animal Feed Science and Technology. 2008;**142**(3-4):317-329

[27] Delgado-Pertíñez M, Gómez-Cabrera A, Garrido A. Predicting the nutritive value of the olive leaf (Olea europaea): Digestibility and chemical composition and in vitro studies. Animal Feed Science and Technology. 2000;**87**(3-4):187-201

[28] Ahmad-Qasem MH et al. Influence of freezing and dehydration of olive leaves (var. Serrana) on extract composition and antioxidant potential. Food Research International. 2013;**50**(1):189-196

[29] Sindi HA. Evidence that supports the antidiabetic, antihypertensive, and

antihyperlipidemic effects of olive (Olea europapaea L.) leaves extract and its active constituents (oleuropein) in human. Journal of Biochemical Technology. 2020;**11**(2):41-45

[30] Cifá D et al. Enhanced yield of oleuropein from olive leaves using ultrasound-assisted extraction. Food Science & Nutrition. 2018;**6**(4):1128-1137. DOI: 10.1002/fsn3.654

[31] Barbaro B et al. Effects of the olive-derived polyphenol oleuropein on human health. International Journal of Molecular Sciences. 2014;**15**(10):18508-18524. DOI: 10.3390/ijms151018508

[32] Şahin S, Bilgin M. Olive tree (Olea europaea L.) leaf as a waste by-product of table olive and olive oil industry: A review. Journal of the Science of Food and Agriculture. 2018;**98**(4):1271-1279. DOI: 10.1002/jsfa.8619

[33] Rocha J, Borges N, Pinho O. Table olives and health: A review. Journal of Nutritional Science. 2020;**9**. DOI: 10.1017/jns.2020.50

[34] Uylaşer V, Yildiz G. The historical development and nutritional importance of olive and olive oil constituted an important part of the Mediterranean diet. Critical Reviews in Food Science and Nutrition. 2014;**54**(8):1092-1101. DOI: 10.1080/10408398.2011.626874

[35] Acar-Tek N, Ağagündüz D. Olive leaf (Olea europaea L. folium): Potential effects on Glycemia and Lipidemia. Annals of Nutrition and Metabolism. 2020;**76**(1):10-15. DOI: 10.1159/000505508

[36] Yateem H, Afaneh I and Al-Rimawi F. Optimum Conditions for Oleuropein Extraction from Olive Leaves. AQU researchers publications. 2014

[37] Ramírez E et al. Oleuropein hydrolysis in natural green olives: Importance of the endogenous enzymes. Food Chemistry. 2016;**206**:204-209. DOI: 10.1016/j.foodchem.2016.03.061

[38] Nekooeian AA, Khalili A, Khosravi MB. Effects of oleuropein in rats with simultaneous type 2 diabetes and renal hypertension: A study of antihypertensive mechanisms. Journal of Asian Natural Products Research. 2014;**16**(9):953-962. DOI: 10.1080/10286020.2014.924510

[39] Rodriguez-Rodriguez R et al. Effects of pomace olive oil-enriched diets on endothelial function of small mesenteric arteries from spontaneously hypertensive rats. British Journal of Nutrition. 2009;**102**(10):1435-1444. DOI: 10.1017/S0007114509990754

[40] Nekooeian AA, Khalili A, Khosravi MB. Oleuropein offers cardioprotection in rats with simultaneous type 2 diabetes and renal hypertension. Indian Journal of Pharmacology. 2014;**46**(4):398. DOI: 10.4103/0253-7613.135951

[41] Medina-Remon A et al. The effect of polyphenol consumption on blood pressure. Mini Reviews in Medicinal Chemistry. 2013;**13**(8):1137-1149. DOI: 10.2174/1389557511313080002

[42] Lubos E, Loscalzo J, Handy DE. Glutathione Peroxidase-1 in Health and Disease: From Molecular Mechanisms to Therapeutic Opportunities. Antioxid Redox Signal. 2011. DOI: 10.1089/ars.2010.3586

[43] Leopold JA, Loscalzo J. Oxidative risk for atherothrombotic cardiovascular disease. Free Radical Biology and Medicine. 2009;**47**(12):1673-1706. DOI: 10.1016/j.freeradbiomed.2009.09.009

[44] Ivanov M et al. Highly potent antioxidant Olea europaea L. leaf extract affects carotid and renal haemodynamics in experimental hypertension: The role of oleuropein. EXCLI Journal. 2018;**17**:29. DOI: 10.17179/excli2017-1002

[45] Charoenprasert S, Mitchell A. Factors influencing phenolic compounds in table olives (Olea europaea). Journal of Agricultural and Food Chemistry. 2012;**60**(29):7081-7095. DOI: 10.1021/jf3017699

[46] Sun W, Frost B, Liu J. Oleuropein, unexpected benefits! Oncotarget. 2017;**8**(11):17409. DOI: 10.18632/oncotarget.15538

[47] Clewell AE et al. A comprehensive toxicological safety assessment of an extract of Olea europaea L. leaves (bonolive™). International Journal of Toxicology. 2016;**35**(2):208-221. DOI: 10.1177/1091581815619764

[48] Guex CG et al. Safety assessment of ethanolic extract of Olea europaea L. leaves after acute and subacute administration to Wistar rats. Regulatory Toxicology and Pharmacology. 2018;**95**:395-399. DOI: 10.1016/j.yrtph.2018.04.013

[49] Omer SA et al. Toxicity of olive leaves (Olea europaea L.) in Wistar albino rats. Asian Journal of Animal and Veterinary. 2012;**7**(11):1175-1182